Deployment Experiences of Guard and Reserve Families

Implications for Support and Retention

Laura Werber Castaneda, Margaret C. Harrell,
Danielle M. Varda, Kimberly Curry Hall,
Megan K. Beckett, Stefanie Stern

Prepared for the Office of the Secretary of Defense

NATIONAL DEFENSE RESEARCH INSTITUTE

The research described in this report was prepared for the Office of the Secretary of Defense (OSD). The research was conducted in the RAND National Defense Research Institute, a federally funded research and development center sponsored by the OSD, the Joint Staff, the Unified Combatant Commands, the Department of the Navy, the Marine Corps, the defense agencies, and the defense Intelligence Community under Contract W74V8H-06-C-0002.

Library of Congress Cataloging-in-Publication Data

Deployment experiences of Guard and Reserve families : implications for support and retention / Laura Werber Castaneda ... [et al.].
    p. cm.
  Includes bibliographical references.
  ISBN 978-0-8330-4573-7 (pbk. : alk. paper)
  1. United States. National Guard Bureau. 2. United States—Armed Forces—Reserves. 3. Deployment (Strategy)—Social aspects—United States. 4. Families of military personnel—United States—Interviews. 5. Families of military personnel—Services for—United States. 6. United States—National Guard—Recruiting, enlistment, etc. 7. United States—Armed Forces—Reserves—Recruiting, enlistmént, etc. I. Castaneda, Laura Werber.

UA42.D45 2008
355.1'293—dc22

                                                            2008044848

Published 2008 by the RAND Corporation
1776 Main Street, P.O. Box 2138, Santa Monica, CA 90407-2138
1200 South Hayes Street, Arlington, VA 22202-5050
4570 Fifth Avenue, Suite 600, Pittsburgh, PA 15213-2665
RAND URL: http://www.rand.org/
To order RAND documents or to obtain additional information, contact
Distribution Services: Telephone: (310) 451-7002;
Fax: (310) 451-6915; Email: order@rand.org

Preface

The use of the Guard and Reserve has steadily increased since the first Gulf War. Additionally, the 2005 National Defense Authorization Act codified the Reserve Component's transition from a strategic reserve to an operational reserve, suggesting that this increased utilization will continue as the Global War on Terror persists. However, demographic differences between active component and reserve component families suggest that the latter may face different issues during deployment and consequently require different types of support.

This monograph is intended as input to the Department of Defense (DoD) as it determines how best to support guard and reserve families. Our research, conducted in 2006, provides an analysis of reserve component families' experiences and perceptions related to deployment. In particular, this monograph focuses on how Army National Guard, Army Reserve, Marine Forces Reserve, and Air Force Reserve personnel and spouses regarded their family's level of readiness for deployment, the problems and positive aspects related to the deployment, their family's ability to cope, and their use of resources. Additional findings pertain to the implications that families' experiences and opinions have for retention intentions, as well as to families' suggestions for improved support.

This monograph should be of interest to military policymakers and the analytical community that studies the Reserve Component or military families, especially researchers focusing on the retention of military personnel. This work should also be of interest to proponents of the Reserve Component and military families, as well as to military service members and their spouses.

This research was sponsored by the Office of the Secretary of Defense (OSD) and conducted within the Forces and Resources Policy Center of the RAND National Defense Research Institute, a federally funded research and development center sponsored by the Office of the Secretary of Defense, the Joint Staff, the Unified Combatant Commands, the Department of the Navy, the Marine Corps, the defense agencies, and the defense Intelligence Community. The principal investigators are Laura Castaneda and Margaret Harrell. Comments are welcome and may be addressed to laura_castaneda@rand.org and margaret_harrell@rand.org.

For more information on RAND's Forces and Resources Policy Center, contact the Director, James Hosek. He can be reached by email at james_hosek@rand.org; by phone at 310-393-0411, extension 7183; or by mail at the RAND Corporation, 1776 Main Street, Santa Monica, California 90407-2138. More information about RAND is available at www.rand.org.

Contents

CHAPTER FOUR
What Problems Do Guard and Reserve Families Report?

CHAPTER FIVE
What Positives Do Guard and Reserve Families Report?

CHAPTER SIX
How Well Do Guard and Reserve Families Cope?

Figures

Tables

Summary

The nation's reliance on the Reserve Component, which includes the Army National Guard, Air National Guard, Army Reserve, Navy Reserve, Air Force Reserve, Marine Forces Reserve, and Coast Guard Reserve, has steadily increased since the first Gulf War in 1990–1991. Over 550,000 reserve component members have been deployed to Operations Enduring Freedom and Iraqi Freedom, and these guardsmen and reservists represent almost 30 percent of all deployments.[1]

This increased dependence on the Reserve Component has implications for reserve families. Although some research has examined the effect of deployment on service members and their families, such research has focused almost exclusively on the Active Component. Because reserve component personnel and their families differ from their active component counterparts demographically, such research may have only limited applicability to reserve component families. For example, reserve component personnel tend to be older than their active component counterparts, and a greater proportion of the Reserve Component is female. Further, guard and reserve families tend to be more geographically dispersed, which may have important implications for how best to support them.

---

[1] From October 1, 2001, to October 31, 2007.

Scope of Research and Methodology

This research addressed family deployment-related issues of concern and interest to the entire Reserve Component. We conducted military family expert interviews that include professionals representing six reserve components; only the Coast Guard Reserve was excluded. However, the interviews with service members and spouses themselves were limited to four of the reserve components: Army Reserve, Army National Guard, Air Force Reserve, and Marine Forces Reserve. Additionally, this research focused on junior and mid-grade enlisted families and on junior officer families. This research scope was determined in conjunction with the research sponsor, reflecting the level of funding available for this research and a focus on personnel who had not already committed to a long military career. This research also focused on guard and reserve families that had experienced at least one deployment outside the continental United States (OCONUS) since 9/11.

The cross-sectional data summarized in this monograph stem primarily from interviews with military family experts, service members, and spouses. Initially, we interviewed, via telephone, individuals identified as experts on the issues concerning reserve component families. This effort included 15 interviews with DoD employees who represented each of the DoD reserve components and the Office of the Secretary of Defense, and also 11 interviews with experts from military advocacy and support organizations. The core of this research is our analysis of the telephone interviews we conducted during the summer of 2006 with 296 service members and 357 spouses of service members, representing 653 guard and reserve families. These interviews consisted of closed-ended questions as well as open-ended questions that were transcribed, reviewed, coded, and analyzed for this research. They provide within this research a rich, qualitative description of the experiences of reserve component families. This summary provides a brief overview of the responses, while the main text of this monograph provides considerably more detail, including exemplary comments as well as an analysis of the interviewee characteristics that help explain differences in comments provided or experiences reported. These characteristics include some of the demographic attributes that differ between reserve component and active component families on the whole, as well as other potentially important factors, such as indicators of maturity, relationship strength, and experience with military life and deployments.

Research Questions

How Ready Are Guard and Reserve Families?

Family readiness is regarded as a critical aspect of preparedness for a service member's active duty service. DoD has stated that "The Department's ability to assist service members and their families to prepare for separations during short and long term deployments is paramount to sustaining mission capabilities and mission readiness" (Office of the Assistant Secretary of Defense for Reserve Affairs, no date). However, how family readiness is defined and measured varies, and some surveys of reserve component service members overlook this subject entirely. This research assessed the meaning of family readiness to both service members and spouses. Overall, three types or components of family readiness were each cited by approximately two-fifths of interview participants: financial readiness, readiness related to household responsibilities, and emotional or mental readiness. Additional, less frequently mentioned aspects of family readiness included those related to legal matters, military resources, and getting a support system in place.

Financial readiness includes an assortment of financial tasks, including saving money in anticipation of a break in pay or in case of emergency, notifying creditors, and both short and long-term financial planning. This was the most frequently cited type of family readiness overall, mentioned by 58 percent of the service members in our study who provided a definition and by 45 percent of the interviewed spouses who provided a definition. Readiness related to household responsibilities includes preparing to handle household responsibilities normally taken care of by the service member, as well as making arrangements related to children. Among those who provided a definition of readiness, this kind of readiness was mentioned by comparable percentages of service members and spouses: 50 percent and 48 percent, respectively. Comments pertaining to emotional or mental readiness included a number of references to "being mentally ready" or having enough time for all family members to "deal with" the fact that the service member will be separated from his or her family for a potentially considerable length of time. Among those who defined family readiness, emotional or mental readiness was mentioned by more than half—approximately 54 percent—of spouses and a significantly smaller proportion—37 percent—of service members.

After asking service members and spouses how they defined family readiness, we then qualitatively assessed how ready they felt their family was for their most recent deployment. Overall, 65 percent of the service members and 60 percent of the spouses in our study indicated that their family was ready or very ready. Approximately one-sixth of both service members and spouses characterized their family as somewhat ready, and approximately one-sixth of both groups characterized their family as not at all ready. Additional analyses not only showed which spouse and service member characteristics helped account for differences in reported family readiness levels, but also demonstrated a strong interrelationship between family readiness and military preparedness. Specifically, we found that service members who said they were well prepared for active duty tended to characterize their family as ready or very ready, while those who believed they were poorly prepared for active duty tended to feel their family was not ready at all. Given the cross-sectional nature of our data, however, we could not determine whether one type of readiness affected the other, or if a third factor, such as an underlying personal attribute, influenced both family readiness and military preparedness.

What Problems Do Guard and Reserve Families Report?

When we asked experts on reserve component family issues about problems that they believed reserve families confront, the majority of experts indicated that guard and reserve families experience the following problems: financial problems, health care issues, emotional or mental problems, and household responsibility issues. We subsequently heard about many of these problems from service members and spouses themselves during our interviews, but to varying degrees. Emotional or mental problems were mentioned most frequently; 39 percent of spouses and 26 percent of service members mentioned such problems. The interviews suggest a range of severity of these problems, from relatively mild sadness and anxiety to more severe emotional or mental difficulties that required medical attention. Problems with household responsibilities were also frequently mentioned by spouses, and almost as frequently mentioned by service members. These comments related to accommodating the demands of family life, including difficulties with child care, household chores, and chauffeuring children. Children's issues were mentioned by 26 percent of spouses and 12 percent of service members. These issues included a range of emotional or mental problems as well as other sacrifices or difficulties experienced by children of deployed service members. While financial/legal problems and health care problems were emphasized by the reserve family experts, they were mentioned by relatively small portions of the service members and spouses interviewed: 15 percent of all interviewees mentioned financial or legal issues, and only about 10 percent mentioned health care problems. Other problems mentioned and discussed in this monograph involve education, employment, and marital strife. Additionally, 29 percent of service members (albeit only 14 percent of spouses) reported that their family had experienced no problems stemming from deployment.

What Positives Do Guard and Reserve Families Report?

The majority of guard and reserve families do experience some positives as a result of activation and deployment. Twenty percent of service members and 29 percent of spouses mentioned increased family closeness as a result of the deployment experience. Twenty-six percent of service members and 20 percent of spouses mentioned financial gain as a positive. Twenty-four percent of spouses and 15 percent of service members mentioned some combination of patriotism, pride, and civic responsibility as a positive aspect. Roughly 20 percent of interviewees mentioned that spouses or families at home felt an increase in independence, confidence, or resilience as result of the deployment. Although the majority of all interviewees reported a positive aspect of deployment, 20 percent of service members and 13 percent of spouses indicated that their family had not experienced any positives as a result of the deployment.

How Well Do Guard and Reserve Families Cope?

Because prior research had focused on the coping ability of families, and despite the potential ambiguity of this concept, we asked interviewees what coping meant for their family and how well they had coped. A sizable minority—37 percent of service members and 29 percent of spouses—were unable to provide any definition of what they meant by coping. The definitions that were provided included the notions of coping emotionally and also coping with household responsibilities, but neither was mentioned by a majority. Despite the absence of a consistent, predominant definition, almost all respondents were able to assess how well their family had coped with deployment, and the majority (63 percent of service members and 62 percent of spouses) said that they or their family coped well or very well.

What Resources Do Guard and Reserve Families Use During Deployment?

In addition to considering the problems and positives, this research also examined the resources to which families turn for support during deployment and why families may not be accessing resources. Our interviews included questions about both the military resources and the informal, nonmilitary resources that families used. Our findings indicate that most of the guard and reserve families we interviewed used some type of resource during their most recent deployment experience. The most frequently cited military resources included TRICARE and family support organizations (such as Family Readiness Groups or Key Volunteer Networks). Military OneSource was a distant third resource, in terms of frequency of mention. Among the nonmilitary resources, the most frequently mentioned were extended family, religious organizations, and friends and neighbors. Across both military and nonmilitary resources, only extended family was cited by a majority of interviewees (among the spouses) as a resource they used during deployment.

How Do Guard and Reserve Families' Retention Plans Differ?

This research included analysis of the service member's intent to remain in the reserve component until retirement eligibility and the spouse's opinion toward his military career. Both service members and spouses were asked versions of both of these questions. For example, service members were asked how they perceived their spouse's attitude toward their military career. In addition, service members were asked to evaluate how their most recent activation affected their career plans. Spouses and service members responded similarly to the question about career intentions; just over half of each indicated plans for the service member to remain in the Guard or Reserve until retirement eligibility. Forty-one percent of spouses and 42 percent of service members indicated plans to leave prior to retirement eligibility. Thirty-eight percent of service members said their most recent activation had no influence on their career plans, while comparable percentages of service members indicated it either increased their desire to stay or increased their desire to leave (30 percent and 32 percent, respectively). Fifty-eight percent of spouses interviewed favored their service member staying in the Guard or Reserve whereas significantly fewer service members—35 percent— believed that their spouse favored their staying.[2]

Similar to the other topics highlighted in this summary, a number of patterns based on demographic attributes and other differing characteristics help to explain variation in reported retention intentions, and they are detailed in the body of this monograph. In addition, we found that family readiness, many of the problems and positives cited by families, and family coping all had implications for retention and, consequently, military effectiveness. Specifically, those who described their family as ready or very ready for the deployment and those who believed their family coped well tended to have a preference for staying. The same was true for those who mentioned one of the major positive aspects of deployment: financial gain, increased family closeness, or patriotism and pride. Conversely, many of the most frequently men-

---

[2]   Interviewed service members and spouses were from different households, so it is unclear whether the individuals married to the service members in our study actually had less favorable views than the spouses interviewed.

tioned problems had negative implications for retention. Those who cited problems related to emotional or mental concerns, employment, education, marital issues, or health care all were more likely to express a preference for leaving.

Conclusion

This research features a rich description of both the problems faced and the positive aspects enjoyed by guard and reserve families as a result of deployment. Our detailed analysis of characteristics that explain which families tend to experience particular types of problems or positives should guide policymakers as they endeavor to support reserve component families. In short, we found that the majority of families mention a deployment-related problem, yet the *kinds* of problems and the *types* of families associated with each problem both differed. The majority of families also cited a positive aspect of deployment, and, as with problems, the characteristics of the families likely to report different positives varied. It should be noted, however, that our exploratory analysis, based on cross-sectional data, did not permit us to address causality or to control for interactions between different characteristics. Thus, we are unable to say, for example, whether family readiness has a direct effect on individual military preparedness, or whether age, pay grade, and marriage length—three potentially interrelated attributes—each have a separate influence on the problems and positives experienced.

It is important that policymakers and those organizations chartered to support military families understand the problems encountered and the positives enjoyed by military families, for several reasons. First, DoD has committed to ensuring and promoting general family well-being as part of a "new social compact" that recognizes the tremendous sacrifice of military families (Office of the Deputy Under Secretary of Defense for Military Community and Family Policy [MCFP], 2002). Second, not only is family readiness viewed as critical to mission success, but quality-of-life issues in general are regarded by DoD as inseparable from overall combat readiness (Myers, 2004). Finally, our analysis indicates a relationship between families' problems and positives and military outcomes, including readiness and retention intentions, that affect DoD's ability to satisfy the military mission.

While many of the problems and the positives merit short-term attention and the allocation of support resources, our findings suggest that successful family support should be assessed, perhaps even primarily assessed, in terms of family readiness, family coping, and retention intentions, as these are measures of military manpower and family-related outcomes that can guide long-term management of reserve component personnel. Unlike the problems and positives families identified, there were common patterns across these three interrelated metrics in terms of who tended to respond in ways with favorable implications for family well-being and military effectiveness: being ready or very ready for deployment, coping well or very well, and expressing a preference for staying. For example, in general, more mature interviewees, those in stronger relationships (as suggested by marriage length), and those with prior military experience were more likely to be ready for deployment and to indicate a preference for staying in the Guard or Reserve. In a similar vein, spouses who had children, were married longer, or were married to service members with a record of prior active duty service also were more likely to report that their family had coped well or very well.

Recommendations

The recommendations were informed by spouse and service member suggestions for improvement, but they neither adopt all those suggestions nor are limited to interviewees' comments. We view these recommendations as constructive steps, but we cannot estimate the result of these changes or their cost-effectiveness without further analysis. In some instances, DoD policymakers, including those within OSD and the services, have begun to implement policies and programs consistent with these recommendations. Our research suggests that such actions may prove effective, and our recommendations underscore their importance.

These recommendations are divided into those related to activation and deployment personnel practices, families' expectations and perceptions, support of and information for families, and measurement of important constructs and outcomes.

Activation and Deployment Personnel Practices

Pursue predictable mobilization in terms of both the length of deployment and the amount of notice.

Ensure that any notice sufficient for service members and families to prepare for deployment is also sufficient for the military to prepare to accommodate the entire family. There should not, for example, be delays in receiving pay for guardsmen and reservists.

Limit the average length of mobilization. Our research suggests that spouses and service members experiencing longer deployments, particularly those one year or longer, were more likely to cope poorly with deployment and to express a preference for leaving the military. This recommendation is consistent with and emphasizes the significance of announced intentions to limit guard and reserve mobilizations to one year.

In a similar vein, DoD should **reduce the use of cross-leveling for reserve component personnel**. Our findings suggest the cross-leveling (deploying individuals with units other than their usual drill unit) may have negative implications for family support. DoD has announced efforts to limit this personnel practice that should have favorable implications for guard and reserve families.

Perceptions and Expectations

Ensure that family expectations are consistent with the DoD vision of a Reserve Component that is both operational and strategic. Service members and families should recognize that they are likely to begin a new deployment every six years, and that some service members may be tapped to serve more frequently.

Recognize that family perceptions are sometimes more important than actual experiences. We found this to be the case with amount of activation notice, where the perceived adequacy of the notice received appeared to be a more compelling influence than the actual amount.

Recognize that families focus on "boots away from home" and not "boots on the ground."

Emphasize the positives of activation and deployment. Consistent with prior research, many of our interviewees experienced an increase in income during their deployment, and some of these financial gains were either unanticipated by the service member, or the service member felt that he or she was unusual in enjoying financial gain.

Support of and Information for Families

Increase levels of readiness among not-yet-activated families.

Know how to find families. DoD should improve the centralized data about families to ensure that both notice and information are received in a timely manner.

Seek ways to make deployment-phased and "on-demand" information available to families. Given that families continue to ask for more and better information, but also criticize the pre-deployment deluge of information, **it is important to tailor both the content and amount of material provided to their needs**. Pre-deployment briefings might be sufficient for some spouses, but they might appear to be a "firehose" of information for spouses unfamiliar with deployments. Focused and intensive workshops might be helpful to some spouses, while others may feel that information from centralized Web sites is sufficient.

Explore ways to connect families to one another, including families that live near one another but represent different units or reserve components.

Bear in mind the limited capacity and capabilities of volunteer-based resources, either military or nonmilitary. Many family support organizations, such as Family Readiness Groups, and local community support, such as VFW organizations, depend heavily on volunteers. DoD should recognize both the strengths and the limitations of these organizations and plan accordingly.

Consistent with this, and given the reliance that our families reported on nonmilitary resources, **seek ways to improve awareness** of, and support or partner with, local and community resources for families.

Recognize that different kinds of families confront different issues during deployment and **tailor efforts to avoid and mitigate deployment-related problems.**

Recognize that, just as the problems experienced by families vary, so do the severity and consequences of problems.

Consider not only how to help those families that are struggling, but also how to **reinforce and learn from those families who appear to proceed through the deployment cycle with fewer problems.**

Measurement of Key Constructs and Outcomes

Recognize that family readiness and coping are multifaceted constructs and develop measures accordingly. Given the importance of family readiness and coping to outcomes such as retention intentions, metrics should be developed that take into account their key dimensions, such as emotional or mental aspects for some families and household responsibilities for others.

Recognize that service members and spouses may provide different assessments of the same deployment experience and that data collection efforts that focus exclusively on either population are inherently limited.

Use metrics to consider both the short-term and long-term effectiveness of family support.

Acknowledgments

The authors thank the offices of the Assistant Secretary of Defense for Reserve Affairs and the Deputy Under Secretary of Defense for Military Community and Family Policy for their support. Specifically, we thank project monitors James Scott and Aggie Byers from these offices. We appreciate the data provided by Virginia Hyland of Reserve Affairs, as well as those who provided data from the reserve components.

We value the thoughtful input provided by the guard and reserve family experts that agreed to be interviewed for this research, including the project monitors listed above as well as Colonel Anthony Baker, Sr., and Lieutenant Colonel Richard Flynn (National Guard Bureau); Command Master Chief Boyd Briggs (Navy Reserve); Connie Bryant (Marine Corps Family Team Building); Lani Burnett (Reserve Enlisted Association); Rene Campos (Military Officers Association of America); David Davidson (Naval Enlisted Reserve Association); Colonel Dick Esau (Marine Corps Reserve Association); Bud Haney, Jim Rowoldt, and John McNeil (Veterans of Foreign Wars); Captain Marshall Hanson (Reserve Officers Association); Delores Johnson (Army Community Service); Sylvia Kidd (Association of the U.S. Army); Brenda Liston (Air Force Family Programs); Elizabeth McNeil Moore (Army Reserve Family Program); Ike Puzon (Naval Reserve Association); Joyce Raezer and Kathy Moakler (National Military Family Association); Betty Schuster (Air Force Reserve Command Headquarters); Major Sheila Speegle (Marine Corps Community Services); Kathy Stokoe (Navy Reserve); Michele Traficante (National Guard Association of the United States); Command Master Chief Richard Trimmer (Regional Command, Navy Region SE); and Ann Yates (Enlisted Association of the National Guard of the United States).

We are appreciative of Chintan Turakhia and Dean Williams, who led the team from Abt SRBI, which conducted many of the telephone interviews.

Additionally, we appreciate the constructive reviews provided by RAND colleagues David Kennedy and Terri Tanielian.

Finally, we note that we could not have completed this work without the candid participation of many service members and spouses who dedicated their time to answer our questions and share their experiences in confidential interviews.

Abbreviations

9/11	September 11, 2001
CNGR	Commission on the National Guard and Reserve
DMDC	Defense Manpower Data Center
DoD	Department of Defense
FRG	Family Readiness Group
GWOT	Global War on Terror
MCFP	Office of the Deputy Under Secretary of Defense for Military Community and Family Policy
OCONUS	outside the continental United States
OSD	Office of the Secretary of Defense
PTSD	post-traumatic stress disorder
RA	Office of the Assistant Secretary of Defense for Reserve Affairs
RCS	Report Control Symbol
VA	Department of Veterans Affairs
VFW	Veterans of Foreign Wars

CHAPTER ONE

Introduction

The nation's reliance on the Reserve Component, which includes the Army National Guard, Air National Guard, Army Reserve, Navy Reserve, Air Force Reserve, Marine Forces Reserve, and Coast Guard Reserve, has steadily increased since the first Gulf War in 1990–1991. As noted in the Commission on the National Guard and Reserve's (CNGR's) March 2007 report to Congress, almost 250,000 reserve component personnel were involuntarily activated for Operations Desert Shield and Desert Storm. Involuntary activations persisted throughout the 1990s as guard and reserve personnel conducted missions in support of operations in Haiti, Bosnia, and Kosovo. In total, more than 550,000 guardsmen and reservists have been deployed to Operations Enduring Freedom and Iraqi Freedom. Reserve component members made up more than 40 percent of the troops in Iraq in 2004 (CNGR, 2007), and in total have represented almost 30 percent of all deployed personnel (Defense Manpower Data Center [DMDC], 2007).[1]

Although this number represents a decline since 2004, the National Defense Authorization Act for Fiscal Year 2005 suggested that the utilization of the various reserve components would continue at a high rate. Specifically, Congress modified the stated purpose of the reserve components, and in doing so formally acknowledged the components' evolution from a strategic reserve into an operational reserve. Instead of being relied on exclusively as a source of personnel for major

---

[1] From October 1, 2001, to October 31, 2007.

wars, the reserve components were also officially tasked with contributing to day-to-day military operations. This additional role for the Reserve Component implies that large numbers of guard and reserve personnel will be called to active duty in the coming years.

This increased and ongoing reliance on the Reserve Component also has implications for reserve families. David Chu, Under Secretary of Defense for Personnel and Readiness, explained in his March 2006 statement to the CNGR that these families may experience hardships and challenges stemming from military service, in particular when the reservist or guardsman is away from home for a significant length of time due to activation or deployment. Although a small body of research (e.g., Haas, Pazdernik, and Olsen, 2005; Hosek, Kavanagh, and Miller, 2006) has examined how deployments affect service members and their families, the focus has been almost entirely on the Active Component. Such findings may be less applicable to reserve component families, because they differ from active component families in important ways. For instance, the Active Component and Reserve Component vary demographically. Some of their differences are summarized in Table 1.1, which uses data from Office of the Deputy Under Secretary of Defense for Military Community and Family Policy's (MCFP's) *2005 Demographics Profile of the Military Community* to compare the Active and Reserve Components.[2] Perhaps most notably, both reserve component personnel and their spouses tend to be older than their active component counterparts. Fewer reserve component personnel are married, but the reserve component has larger percentages of both female service members and male spouses. In addition, although the percentage of personnel with children is the same across the two components, measures of the average number of children under age 18 per

---

[2]   The Active Component includes the Army, Navy, Marine Corps, and Air Force, while the Reserve Component includes the seven reserve components: Army National Guard, Army Reserve, Naval Reserve, Marine Forces Reserve, Air National Guard, Air Force Reserve, and Coast Guard Reserve. The Reserve Component figures within this monograph exclude the Coast Guard Reserve and are based on Selected Reserve personnel numbers; personnel in the Individual Ready Reserve or the Inactive National Guard are not part of any calculations in this monograph.

Table 1.1
Demographic Comparison of Active and Reserve Components

	Overall Active Component	Overall Reserve Component[a]
Female service members	15%	17%
Service members age 25 or younger	47%	31%
Average age for officers	34.6	40.5
Average age for enlisted personnel	27.1	31.8
Married service members	55%	51%
Male spouses	7%	11%
Spouses age 25 or younger	29%	12%
Service members with children[b]	43%	43%
Average number of children under 18, per parent service member	1.9	1.7

SOURCE: MCFP, 2005.

[a] The Reserve Component category consists of six of the reserve components; it excludes the Coast Guard Reserve.

[b] The definition of children includes dependents age 23 or younger and dependents enrolled as full-time students.

parent show that active component personnel tend to have younger dependents.

These demographic differences suggest that research pertaining to support of active component families and the actual programs and services available to those families may be less appropriate for reserve component service members and their families. It is possible, for example, that older service members, older spouses, and older children may have fewer or different needs than younger ones, or that families of female personnel experience deployments differently than those of male personnel. Further, as Chu noted to the CNGR, "[M]any families of National Guard and Reserve members do not live close to a military installation where many of the traditional family support activities are located" (Chu, 2006, p. 17). In his remarks, Chu discussed establishing family support centers around the country and increasing the use of information technology, such as the World Wide Web, in response to the typical reserve component family's distance from military installations, but it is unclear whether those resources adequately compensate for the standard services available on military installations that are readily accessible to active component families.

The small amount of research on how deployments affect families (particularly guard and reserve families), along with current and future plans for intensive utilization of reserve component personnel, motivated the current study. Specifically, we sought to assess the issues faced by guard and reserve families during activation and deployment, explore their perceptions and use of family support resources, and analyze how these experiences and perceptions influence their planned retention. In this document, we summarize our findings, recommend policy changes as necessary to support and retain these families, and identify strategies for future data collection efforts.

Scope of Research

This research addressed family deployment-related issues of concern and interest to the entire Reserve Component. We conducted expert interviews that include professionals representing six reserve components; the Coast Guard Reserve was not included. However, the interviews with service members and spouses themselves were limited to four of the reserve components: Army Reserve, Army National Guard, Air Force Reserve, and Marine Forces Reserve. Additionally, this research focused on junior and mid-grade enlisted families and on junior officer families. This research scope was determined in conjunction with the research sponsor, reflecting a focus on individuals who had not already committed to a long military career and the level of funding available for this research. This research also focused on guard and reserve families that had experienced at least one deployment outside the continental United States (OCONUS) since 9/11.

Methodology

Expert Interviews

The data summarized in this monograph stem primarily from interviews with military family experts, service members, and spouses. Initially, the research team interviewed, via telephone, individuals whom we identified, or that were identified by our research sponsor, as experts on the issues concerning reserve component families. In total, we conducted 15 interviews with military family experts employed by the Department of Defense (DoD) to represent each of the DoD reserve components and the Office of the Secretary of Defense (OSD), and also 11 interviews with experts from advocacy and support organizations, including the National Military Family Association, the Military Officers Association of America, and the Reserve Enlisted Association. A full list of the organizations represented in these expert interviews, which were conducted winter 2006, is provided in Appendix A. Topics covered during these interviews included

- problems and positive aspects of deployment
- differences between reserve component and active component families as well as between families from different reserve components
- the effect of geographic location on family problems and support
- resources available to reserve component families
- organizational challenges to supporting these families
- suggestions for improvement.

The military family expert interview protocol that we used is also included in Appendix A. These interviews were transcribed and analyzed using an inductive process, and excerpts from these interviews appear in this monograph.

Service Member and Spouse Interviews

Sample and Sampling Issues. The core of this research is our analysis of the telephone interviews we conducted during the summer months of 2006 with 296 service members and 357 spouses of service members. As discussed previously in the "Scope of Research" section, these interviews included personnel and spouses from four of the six DoD reserve components. The spouse and service member interview protocols covered many of the same topics addressed in the expert interviews, such as problems and positive aspects related to deployment, resources available to guard and reserve families, and suggestions for improvement, along with additional questions pertaining to such issues as the effect of activation notice on the family, family readiness, and family coping. Demographic attributes were also collected during the course of the interview, and a series of multiple-choice questions about the family's current financial situation, the service member's military career intentions, and the spouse's opinion regarding the service member's military career intentions were posed throughout the interview. Representative spouse and service member protocols are shown in Appendix B. The actual protocol used varied somewhat for each interviewee because the protocols were dynamic, allowing for different paths of questioning to be pursued depending on the interviewee's response to an earlier question.

To reach service members and spouses of service members, from each of the included reserve components we obtained contact data for service members who both had deployed OCONUS since 9/11 and had dependents. Similar data were also obtained for spouses of service members who met the same criteria. Our original sampling plan had been to interview 90 service members and 90 spouses from each of the four DoD reserve components, for a total of 180 households per component. We were interested in service members who had recently demobilized and spouses of service members who had either recently demobilized or who were still deployed. Our past experience contacting military families suggested a minimum number of service members and spouses for whom we required contact information such that we could achieve our desired number of interviews via random sampling,[3] but ultimately we were limited by the quantity of family data available. In some instances, the data we received included data available for all service members who had deployed OCONUS since 9/11, but we still had difficulty reaching sufficient participants. With other components, we received data representing some—but not all—of the service members and spouses who experienced a deployment since 9/11. Such was the case with the Army National Guard, which furnished us sufficient data by providing the contact information for service members and spouses in a few states that it chose in consultation with the sponsoring office of this research. To summarize, all four reserve components furnished contact information for spouses and service members that met our criteria, but not every reserve component provided us with contact information for *all* the individuals who met our requirements.

Using the data provided, we constructed a demobilized service member population and a spouse population that included spouses of both demobilized and deployed service members. We mailed to everyone in those populations introductory letters that both described the study and explained that the addressee may be randomly selected to participate in a telephone interview. We stratified those samples by the pay grade groupings indicated in Tables 1.2 and 1.3, and then randomly selected individuals from each of those pay grade categories for telephone interviews.[4]

This effort was hampered by the number of incorrect addresses and telephone numbers for service members and spouses. We tracked not only the number of introductory letters we sent, but also the number of

---

[3]   This minimum number was four to five times the number of desired interviews in a particular pay grade category. For example, we sought to conduct 27 spouse interviews in the E1–E4 pay grade category, and thus requested contact information for at least 135 spouses (5 times 27) married to a service member deployed OCONUS at least once since 9/11. Some reserve components provided us with information that exceeded our request, while others closely adhered to it.

[4]   The final sample included a small number of service members who had recently been promoted to E-7 or O-4, and spouses whose service member had recently received a similar promotion. Because of the difficulty obtaining a sufficient sample, we chose to include this small number of individuals. They are included in the E-6 or O-3 pay grade categories in Tables 1.2 and 1.3, and were subsequently analyzed in those groupings.

Table 1.2
Breakdown of Personnel Interviewed, by Reserve Component and Pay
Grade Category

Pay Grade	Original Quotas	Army National Guard	Army Reserve	Marine Forces Reserve	Air Force Reserve
O-1 to O-3	23	27	22	21	11
E-1 to E-4	27	32	11	17	9
E-5 to E-6	40	45	41	20	40
Totals	90	104	74	58	60

SOURCE: 2006 RAND Guard and Reserve Family Interviews.

NOTES: Gray shading denotes original interview quotas; the remainder of the table indicates interviews performed. N=296 service members.

Table 1.3
Breakdown of Spouses Interviewed, by Reserve Component and Pay
Grade Category

Pay Grade	Original Quotas	Army National Guard	Army Reserve	Marine Forces Reserve	Air Force Reserve
O-1 to O-3	23	23	23	24	23
E-1 to E-4	27	35	27	19	9
E-5 to E-6	40	44	39	40	51
Totals	90	102	89	83	83

SOURCE: 2006 RAND Guard and Reserve Family Interviews.

NOTES: Gray shading denotes original interview quotas; remainder of table indicates interviews performed. N=357 spouses.

letters returned for incorrect addresses. A detailed breakdown of return rates by component and pay grade category is provided in Appendix B. For service members, the percentage of letters returned ranged from 5 to 17 percent, depending on reserve component. Accurate contact data were particularly a problem for Army Reserve personnel and for those in the junior enlisted ranks; 17 percent of letters addressed to Army Reserve personnel were returned, as were 13 percent of those sent to junior enlisted personnel from all four reserve components. For spouses, the percentage of letters returned varied from 5 to 7 percent by reserve component. Again, data related to the junior enlisted ranks proved to be more problematic, with 8 percent of all letters sent to spouses of junior enlisted personnel returned. On the whole, the contact information for spouses was more reliable than that for service members.

Additionally, we found that many of the telephone numbers provided by the reserve components were mobile telephone numbers, which in accordance with the Telephone Consumer Protection Act (TCPA) cannot be autodialed, as was necessary for random number sampling. Similar to our procedure for the letters returned due to incorrect addresses, we assessed the number of unusable mobile telephone numbers. A detailed breakdown of mobile telephone numbers by component and pay grade category is provided in Appendix B. For the service member population, the average number of mobile telephone numbers varied from 17 to 32 percent by reserve component, while for the spouse population, comparable figures range from 17 to 35 percent by reserve component. In both populations, there were high rates of mobile telephone numbers across the pay grade categories, but the numbers were especially high among junior enlisted personnel and spouses: 32 percent of junior enlisted personnel in the service member population and 34 percent of junior enlisted spouses in the spouse population had only mobile telephone numbers in the contact information provided.

Despite efforts to contact prospective interview participants via their mobile telephone numbers to obtain landline telephone numbers that could be autodialed, we were required to restructure our sampling to accommodate the limitations of the data. Further, some interviews were conducted with service members who were incorrectly listed as having dependents, and they were subsequently excluded from our analysis.

Interview Disposition. In general, service members and spouses were typically willing to be interviewed. Overall, 9 percent of spouses declined to participate, ranging from 6 percent of Army National Guard spouses to 16 percent of Marine Forces Reserve spouses. Eight percent of service members opted not to participate, ranging from 6 percent of Marine Corps reservists to 9 percent of Army reservists. The interviews were relatively short, averaging 23 minutes in length for spouses and 22 minutes for service members. Roughly 80 percent of all interviews were completed during the evening, and about 75 percent of them were conducted during the week (Monday through Friday). Interviewee characteristics, along with how they compare with the overall population of service members and spouses in their respective reserve components, are discussed in Chapter Two.

Recording, Coding, and Analysis of Service Member and Spouse Interviews

During the interviews with service members and spouses, the responses to closed-ended questions were entered directly into a computerized survey system. The audio responses to selected open-ended questions were recorded and later transcribed, and the responses to other open-ended questions were typed by the interviewer immediately following that response.

The data from the closed-ended questions were analyzed using conventional statistical methods. The data from the open-ended questions were transcribed and later coded and analyzed for prevalent themes and notable patterns across different types of interviewees. The coded results were then statistically analyzed, and the results of these statistical tests provide the empirical foundation for Chapters Three through Nine. In all cases, findings reported are statistically significant at p<.10, and in many instances, specific percentages are reported to illustrate both how frequently a particular response was given and the magnitude of the difference between groups (e.g., by reserve component). However, the reader should bear in mind that just as survey data have a margin of error, so too do qualitative data. Accordingly, greater attention should be given to the nature of differences and their relative magnitude rather than to precise percentages or percentage point differences.

Additional information about our coding and analysis is provided in Appendix B, including a description of the coding procedures and a list of codes used to tag interview text.

Limitations of Interview Data

The goal of this exploratory study was to generate insights regarding the experiences of reserve component families that would inform policy as well as the development of programs and services intended to support these families. Given the lack of research on this topic, an inductive approach was employed. Accordingly, the study emphasizes rich, descriptive findings present within the interview data. Although the statistically significant findings reported herein may be applicable to reserve component personnel and spouses with characteristics similar to those who participated in our interviews, additional research is warranted to determine the extent to which these findings are generalizable to *all* families in the Army National Guard, Army Reserve, Marine Forces Reserve, and Air Force Reserve. This is especially important to note, given the differences in how the reserve components compiled contact information for the study. Although we used stratified random sampling to select interview subjects, we do not know how all the reserve components developed the lists of potential study participants that we used in our initial mailing. Finally, our analysis was not only exploratory but also based on cross-sectional data, and consequently did not permit us to address causality. Thus, we are unable to conclude whether a relationship between two measures or factors is based on the effect one has on the other, or whether a third measure is affecting both of the measures under consideration.

Organization of This Monograph

Chapter Two provides additional details on how reserve component families differ from active component families, describes the spouse and service member portions of our interview sample, and discusses the extent to which our interview sample resembles the larger population of service members and spouses by reserve component. Chapter Three analyzes guard and reserve family readiness for deployment. The next two chapters discuss the problems encountered and the positives incurred by families due to deployment. Chapter Six addresses how families coped with deployments, and Chapter Seven describes the resources that families turn to during deployments. Chapter Eight analyzes the implications that our findings have for retention. Chapter Nine describes the suggestions that interview participants made to improve the support provided to guard and reserve families. Chapter Ten includes our research conclusions and recommendations. This monograph is supplemented by two appendices. Appendix A includes more information about the organizations represented in the military family expert interviews and the protocol used for those interviews. Appendix B includes the introductory letter and the interview protocols used for the service member and spouse interviews, as well as more information about those interviews and our analysis of the interview data.

CHAPTER TWO

What Are the Characteristics of Guard and Reserve Families?

In the preceding chapter, we noted that, as a whole, reserve component families look different from active component families in several ways. In this chapter, we expand on that premise by comparing and contrasting the four reserve components included in our study with their active component counterparts, and with one another. We draw on our military family expert interviews to provide additional insights to this description of guard and reserve families, with emphasis on their potentially unique characteristics and concerns. Finally, we describe our interview participants in terms of how they compare with the overall reserve components from which the interview sample was drawn, as well as with respect to additional characteristics that we considered during our analysis.

Component-Level Comparisons

Table 2.1 provides additional detail on the demographic comparisons discussed in Chapter One. Specifically, rather than comparing the overall Active Component with the entire Reserve Component, a breakdown by individual active and reserve components with respect to gender, age, and family-related measures is provided. This reveals some notable differences between individual active and reserve components that were not apparent with the higher level, aggregate comparison based on the four active components and six DoD reserve components.

Table 2.1
Detailed Demographic Comparison of Active and Reserve Components

	Army	Army National Guard	Army Reserve	Marine Corps	Marine Forces Reserve	Air Force	Air Force Reserve
Female service members	14%	13%	23%	6%	5%	20%	24%
Service members age 25 or younger	46%	37%	34%	66%	66%	39%	16%
Average age for officers	34.6	38.6	41.7	33.4	40	34.3	41.8
Average age for enlisted	27.1	30.9	30.9	24.1	24.4	28.4	35.9
Married service members	54%	49%	48%	45%	31%	61%	59%
Male spouses	7%	7%	14%	2%	3%	9%	16%
Spouses age 25 or younger	31%	16%	12%	42%	30%	24%	6%
Service members with children[a]	46%	42%	41%	30%	21%	46%	51%
Average number of children under 18, per service member parent	1.9	1.8	1.7	1.8	1.8	1.8	1.6
Single-parent service members	7%	8%	9%	3%	3%	5%	9%

SOURCE: MCFP, 2005.

NOTE: Gray shading denotes reserve components.

[a] The definition of children includes dependents age 23 or younger and dependents enrolled as full-time students.

For example, while the Reserve Component has a greater proportion of female service members than does the Active Component, it is a relatively small difference of 2 percentage points. A comparison of the Army Reserve directly with the active component Army, however, shows a difference of 9 percentage points. The Air Force Reserve also has a higher proportion of female personnel than does its active component counterpart: 24 percent of Air Force reservists are women, compared with 20 percent of active component Air Force personnel. The experts we spoke with felt that the larger proportion of women serving in the reserve components was one notable way in which the reserve components differed from the active components. The following comments from military family experts demonstrate how this may have potential implications for family support:

> It is always difficult to have a parent away, but having the mother away is usually tougher. A lot of spouses will have child care issues because they often have no backup. (25: Non-DoD military family expert)[1]

> There are issues specific to males who are the spouse [when] the service member is the wife. They often do not feel connected to the family readiness services offered for spouses, and they often have a difficult time getting involved in the resources available. (18: Non-DoD military family expert)

> Generally, when men are the stay-behind spouses, they are probably less adept at child care and probably less willing to give up employment to stay home and care for children. They need our help with child care to help keep their job. Men probably need more help in that regard. (6: DoD military family expert)

These issues may be especially of concern for the Air Force Reserve and Army Reserve. Approximately one-fourth of each force is female, and

---

[1]   After each quotation, a unique identifier indicates the interview in which the comment was made. The same identifier is used to denote the same interview throughout the report, but it does not have significance nor can it be used to identify the participants. These numerical identifiers are used to convey the extent to which evidence is present in multiple interviews.

the proportion of female personnel is greater than that for their active component counterparts. The Air Force Reserve and Army Reserve also have larger percentages of male spouses compared with the Air Force and Army, respectively. In contrast, the percentages of female personnel in the Army National Guard and Marine Forces Reserve are not only lower than comparable figures for the Army Reserve and Air Force Reserve, but they are also lower than those for the active component Army and the Marine Corps.

As mentioned in Chapter One, reserve component personnel and their spouses are, on average, older then their counterparts in the active components. This is true both in the comparison of the entire Active Component compared with the entire Reserve Component as well as when individual reserve components are compared directly to their active component counterparts, with the exception of the Marine Forces Reserve. The same proportion of service members in the Marine Corps and the Marine Forces Reserve, 66 percent, were age 25 or younger. In addition, while the proportion of spouses age 25 or younger was lower for the Marine Forces Reserve (30 percent) than for the Marine Corps (42 percent), it was still well above the overall Reserve Component average of 12 percent, higher than the proportion for the Air Force, and comparable to the proportion for the Army. Service member and spouse age can be considered a proxy for maturity, and, consequently, younger individuals may have different needs in terms of family support. Comments made during our military family expert interviews illustrate how younger individuals and families may experience activation and deployment differently than their older peers:

> For younger families, they are just getting started. They haven't really learned how each other operates. They really don't know each other. They haven't learned resiliency skills. The younger couples are more likely to battle with trust/security issues because they don't really know each other. Older couples know each other and can navigate this a little easier. Older reservists also know the reserves better and can teach their spouse to navigate this world. (7: DoD military family expert)

I do think that younger families have problems that come with less experience in being a family in general. Younger families are less inclined to ask for help, and older families are more inclined. (2: DoD military family expert)

While this may be the case for the younger service members and spouses in all the reserve components, it may be a particular concern in the Marine Forces Reserve, given its larger proportion of young families. On the other hand, the Army Reserve, Army National Guard, and Air Force Reserve have a larger contingent of older families, as suggested not only by the smaller proportion of service members and spouses age 25 or younger, but also by the lower average number of children under age 18 (per parent). Our expert interviews also provided insights on what this could mean for reserve component families:

Overall, the Reserve Component tends to be older than the Active Component population. . . . Because they are older, they tend to have more teenagers, which creates a unique set of problems: driving them around and getting them to activities when one parent is now missing. But generally, fear is more prevalent in the younger kids, so families with younger kids have different things to deal with than families with teenagers. (9: DoD military family expert)

For [the Air Force Reserve], it's more about caring for extended family or for children in college (with financial support) compared with younger families with children, who instead deal with day-to-day challenges like child care, single parenting, and blended families. (14: DoD military family expert)

The latter comment raises the additional issues of single parenting and blended families. Although the MCFP's *2005 Demographics Profile of the Military Community* does not contain data related to the many types of blended families, it does include figures for single-parent service members. As shown in Table 2.1, the reserve components generally resemble their active component counterparts with respect to the proportion of service members that are single parents. The largest difference is between the Air Force and the Air Force Reserve, and may explain why this issue was salient for the military family expert who raised it in the above remark. While 5 percent of active component Air Force personnel are single parents, this figure is almost twice as high (9 percent) in the Air Force Reserve.

Interview Sample Level Comparisons

Given the potential influence of gender, age, and family-related characteristics in shaping the issues and challenges faced by reserve families, it is important to note how our interview sample looks with respect to these attributes and to consider the extent to which our interview participants resemble the reserve components from which our sample was drawn. Table 2.2 features a comparison of our interview sample with the four reserve components included in our study. Note that the interview sample includes junior officers (O-1 to O-3), junior enlisted (E-1 to E-4) personnel, and mid-grade enlisted (E-5 to E-6) personnel, while the data from the MCFP 2005 demographics report include all officers and all enlisted personnel, and thus Table 2.2 emphasizes the similarities and differences between our sample and the total component. In addition, service member and spouse data are separate in Table 2.2 because the service members and spouses in our study were not married to one another; they represent different households.

For all four components, the average age for officers is lower in our sample than within the components themselves, which is likely due to the study's inclusion of only those officers in the O-1 to O-3 pay grades. However, although senior enlisted personnel (E-7 to E-9) were not included in the interviews, the mean age for enlisted personnel in our study is greater than the mean age for enlisted personnel in each of the four reserve components. Further, the percentage of service members age 25 or younger is smaller in our interview sample. It appears that our enlisted portion of the service member sample is, on average, older than the reserve components from which it was drawn. The situation for spouses differs somewhat, though. While the spouses of Marine reservists we interviewed were older, on average, than Marine Forces Reserve spouses in general, for the Army National Guard and

Table 2.2
Demographic Comparison of Reserve Components with Interview Sample

	Army National Guard	Army National Guard Sample	Army Reserve	Army Reserve Sample	Marine Forces Reserve	Marine Forces Reserve Sample	Air Force Reserve	Air Force Reserve Sample
Service members		N=104		N=74		N=58		N=60
Female service members	13%	4%	23%	8%	5%	12%	24%	15%
Service members age 25 or younger	37%	8%	34%	8%	66%	31%	16%	8%
Average age for officers	38.6	36.0	41.7	39.1	40.0	35.1	41.8	36.3
Average age for enlisted	30.9	36.0	30.9	33.2	24.4	27.9	35.9	36.6
Single parents	8%	9%	9%	12%	3%	45%	9%	12%
Average number of children under 18, per parent	1.8	2.3	1.7	2.3	1.8	1.7	1.6	1.7
Spouses		N=102		N=89		N=83		N=83
Male spouses	7%	4%	14%	2%	3%	1%	16%	6%
Spouses age 25 or younger	16%	15%	12%	20%	30%	22%	6%	5%
Average number of children under 18, per parent	1.8	2.2	1.7	2.1	1.8	1.9	1.6	2.1

SOURCES: MCFP, 2005, and 2006 RAND Guard and Reserve Family Interviews.

NOTE: Gray shading denotes reserve components.

Air Force Reserve, the spouse average ages in the sample are very similar to the means for those reserve components. In addition, the spouses we interviewed for the Army Reserve tended to be younger, on average, than Army Reserve spouses overall. Interestingly, although our interview sample tends to be older than the reserve components from which it was drawn, figures for the average number of children under 18 are higher in most of the sample than comparable values for the overall components. Only the Marine reservists we interviewed had fewer children under 18, on average. Thus, patterns related to children or parenting in our study may have greater prominence than they would have in the overall reserve components. On a related note, more of the service members we interviewed tend to be single parents, which may be explained in part by our need to interview only service members with dependents. Although the difference between the proportion of Marine reservist single parents in our sample (45 percent) and that of the overall Marine Forces Reserve (3 percent) is rather large, 65 percent of the Marine reservist single parents in our study indicated they did have a significant other. It is possible that they, as well as the other single-parent service members in our study with a domestic partner, may share parenting and household responsibilities with that individual.

Turning our attention toward gender, the interview sample tended to differ from the overall reserve components with respect to both the proportion of female service members and the proportion of male spouses. Within the service member portion of the interview sample, the Army National Guard, Army Reserve, and Air Force Reserve had proportionately fewer female service members than did the Army National Guard, Army Reserve, and Air Force Reserve overall. In the largest such difference, 23 percent of Army reservists were women, but only 8 percent of the Army reservists we interviewed were women. Conversely, our sample of Marine reservists had proportionately more women than did the Marine Forces Reserve overall—12 percent compared with 5 percent. In addition, male spouses appear to be consistently underrepresented in our sample. For all four components, the spouse portion of the interview sample has proportionately fewer men. Again, the largest percentage point difference was present for the Army Reserve: Only 2 percent of the Army Reserve spouses we interviewed were men, while 14 percent of all Army Reserve spouses were men. The relatively small number of female service members and male spouses in our interview sample means fewer gender-related patterns can be discerned than might have been the case with a sample more closely in line with the gender diversity in the reserve components themselves.

Tables 2.3 and 2.4 list additional descriptive information about the interview sample. Table 2.3 pertains to the service member portion of the sample, and Table 2.4 addresses the spouse portion. The measures in these tables are not included in the MCFP's *2005 Demographics Profile of the Military Community*, so comparable statistics for the respective research components are not provided. As suggested by the average number of children figures shown in Table 2.2, the vast majority of both the service members and spouses we interviewed were parents of children under age 18. Consistent with the older age of many of the service members and spouses, relatively few of the individuals were newlyweds (which we define as married for two years or less). Among the service members, figures ranged from 8 percent of Air Force reservists to 29 percent of Marine reservists newly married. A larger proportion of the spouses we interviewed indicated they were in new marriages: As few as 12 percent of Air Force Reserve spouses and as many as 31 percent of Marine Forces Reserve spouses reported they were married for two years or less. This is an important distinction because, as noted earlier, marriage length may factor in to how families experience a deployment and may also serve as a proxy for maturity or relationship strength.

Another potential indicator of maturity, as well as of deployment experience, is prior military experience. In all the interviews, we asked whether the service member had prior active duty experience, and during the spouse interviews, we also inquired about the spouse's own prior military experience. We viewed this as an important characteristic because those with prior military experience, especially prior active duty experience, may be more familiar with the military lifestyle and with the resources available to military families than individuals whose only exposure to the military has been through participation in the Guard or Reserve. Several of the military family experts we interviewed suggested the advantages of active duty experience:

Table 2.3
Additional Interview Sample Characteristics—Service Members

	Army National Guard (N=104)	Army Reserve (N=74)	Marine Forces Reserve (N=58)	Air Force Reserve (N=60)
Parent of children under age 18	78%	81%	85%	70%
Married for two years or less	17%	14%	29%	8%
Prior active duty experience	52%	55%	47%	60%
Repeat OCONUS deployments	4%	10%	22%	37%
Comfortable current family financial situation	68%	64%	60%	70%
Employed either full or part-time	83%	85%	88%	92%
Self-employed	4%	4%	7%	5%
Spouse or significant other employed either full or part-time[a]	69%	77%	73%	66%
Reside within 25 miles of drill unit	32%	32%	15%	40%
Reside at least 100 miles away from drill unit	25%	31%	31%	18%
Reside within 25 miles of nearest installation	37%	36%	43%	53%
Reside at least 100 miles away from nearest installation	14%	27%	12%	12%
Received one week's notice or less	6%	15%	20%	22%
Received one month's notice or less	42%	68%	53%	52%
Most recent deployment one year or longer	94%	84%	14%	8%
Average length of most recent deployment (months)	14	13	8	4

SOURCE: 2006 RAND Guard and Reserve Family Interviews.

[a] The percentage of service members was calculated based on the number of service members with a spouse or significant other.

Table 2.4
Additional Interview Sample Characteristics—Spouses

	Army National Guard (N=102)	Army Reserve (N=89)	Marine Forces Reserve (N=83)	Air Force Reserve (N=83)
Parent of children under age 18	83%	76%	64%	76%
Married for two years or less	14%	24%	31%	12%
Prior military experience	10%	5%	2%	22%
Married to service member with prior active duty experience	52%	47%	64%	71%
Married to service member with repeat OCONUS deployments	32%	18%	47%	57%
Comfortable current family financial situation	67%	68%	81%	74%
Employed either full or part time	66%	61%	74%	74%
Reside within 25 miles of drill unit	24%	26%	21%	45%
Reside at least 100 miles away from drill unit	20%	26%	25%	16%
Reside within 25 miles of nearest installation	40%	42%	42%	61%
Reside at least 100 miles away from nearest installation	10%	15%	7%	2%
Received one week's notice or less	3%	6%	8%	6%
Received one month's notice or less	43%	49%	39%	45%
Most recent deployment one year or longer	79%	73%	18%	10%
Average length of most recent deployment (months)	15	12	8	4

SOURCE: 2006 RAND Guard and Reserve Family Interviews.

We know that it is difficult for a family to just one day be told they can use an installation, and then expect that they will know how to enter the base, get around, and utilize services. They hardly know what the acronyms mean at first, so it's unlikely they would be able to step into this foreign setting and feel comfortable. . . . The active component [families] have the military savvy that I'm referring to, so they simply drive up to an installation and use it and can talk to everyone there with confidence. (9: DoD military family expert)

I think that [guard and reserve] families would comment that they are not able to connect to the military system; spouses just don't know enough about it and are not familiar with even the most obvious characteristics. For example, you might have a two-star general whose wife does not even know what a "rank" is and is unable to identify with this. They're just not tied into the military or what it means to be in the military. (15: DoD military family expert)

Among the service members in our study, the majority of all but those in the Marine Forces Reserve had prior active duty experience. For the spouses, a sizable minority of Air Force Reserve spouses, 22 percent, had prior military experience themselves. Comparable figures for the other components' spouses were much smaller, but the service members they were married to often had prior active duty experience. This was especially true of the Marine Forces Reserve and Air Force Reserve spouses we interviewed.

While those with prior active duty experience likely have prior deployment experience, we directly considered deployment experience as part of the Guard or Reserve as well. Specifically, we asked how many times the service member was deployed OCONUS since 9/11. Tables 2.3 and 2.4 show the proportion of service members and spouses who reported experiencing more than one deployment post-9/11. For service members in our sample, the numbers were relatively small, especially for interviewed Army Guardsmen; only 4 percent of them had been deployed OCONUS more than once since 9/11. Comparable figures for the interviewed spouses were higher (recall that the spouses and service members in our study were not married to one another). In both instances, individuals in our sample affiliated with the Air Force Reserve most frequently had experienced more than one deployment since 9/11—37 percent of Air Force reservists and 57 percent of Air Force Reserve spouses, although, as will be discussed shortly, they tended also to report shorter deployments. Previous deployment experience may be an important indicator of a family's familiarity with deployment, but it is not clear whether those who experienced multiple deployments were involuntarily mobilized for their subsequent tours or had volunteered willingly for them. This distinction may have implications for how families respond to the deployment, regardless of their prior deployment experience.

The next items listed in Tables 2.3 and 2.4 depict the financial situation and employment status of those who participated in our study. During the interviews, we asked spouses and service members to characterize their family's current financial situation using a five-point scale that ranged from "very comfortable and secure" to "in over our heads."[2] For ease of presentation, this scale was collapsed into a three-point scale during our analysis; interviewees either had comfortable family finances, uncomfortable family finances, or were financially neutral (i.e., neither comfortable nor uncomfortable). Accordingly, the percentage of service members and spouses who indicated their family had a comfortable current financial situation, as shown in Tables 2.3 and 2.4, includes those who regarded their family's financial situation as either "very comfortable and secure" or "able to make ends meet without much difficulty." The majority of service members and spouses in our study regarded their current family finances as comfortable, with percentages ranging from a low of 60 percent of Marine reservists to a high of 81 percent of Marine Forces Reserve spouses.

Closely related to the family's current financial situation is the employment status of the service member or spouse at the time of the interview. We asked service members and spouses whether they were employed full-time, employed part-time, not employed but seeking

---

[2] Our question was similar to one used in DMDC surveys, most recently the 2006 Survey of Reserve Component Spouses.

employment, or not employed and not seeking employment. Service members with a spouse or significant other were also asked about the spouse's employment status. Given concerns about the particular challenges that self-employed reservists and guardsmen face when activated, an additional question was posed to the service members in our study to determine whether they were self-employed. While most service members were employed either full-time or part-time, with figures ranging from 83 percent of Army guardsmen to 92 percent of Air Force reservists employed, very few service members were self-employed. The small number of self-employed guardsmen and reservists in our study prevented us from systematically examining the unique issues these service members encounter while activated. With respect to spouse employment, the majority of interviewed spouses reported that they were working either full-time or part-time, as was the case for the spouses and significant others of the interviewed service members. There were some differences between the two portions of the sample, however. For example, while 61 percent of Army Reserve spouses that we interviewed were employed, a greater proportion of married service members that we interviewed—77 percent—reported that their spouse was employed.

During our interviews with military family experts, they frequently discussed reserve component families' distance from military installations and resources, identifying that as both a key difference from active component families and a source of challenges for reserve component families. The following comments are representative of their remarks:

> Active families typically have a built-in support structure. That is, they are close to a base and all its services. Reservists can be 40, 60, up to 200 miles (or more) away from the closest installation. There is not a system that makes it as easy [for reserve component families] as for the active families to get support from the military. Not only are services hard to come by, but often the spouse is left behind. They might not know anyone else in the same situation as themselves. (16: Non-DoD military family expert)

Active duty families know to go to an installation [for resources], and there are organizations around an installation engaged in supporting military families. Often reserve families are the only family in their community that is a military family. (4: DoD military family expert)

Reserve families are scattered. Any reserve unit can have members that are miles away or close to the drilling site. When they do go to drill, after the weekend they go back home. The active component [families] cling to each other more readily, especially if they are living on a base or on post, so their community activities are together. (22: Non-DoD military family expert)

The greatest obstacle is that the majority of [reserve] families do not live near an active duty base, which provides morale, welfare, and support to families. Reserve families don't have that same safety net or network. (26: Non-DoD military family expert)

Accordingly, in our interviews with spouses and service members, we asked about the distance of their residence from both the drill unit and the nearest military installation. Tables 2.3 and 2.4 show the proportion of families who lived within 25 miles of the drill unit or the closest military installation. For example, as few as 15 percent of Marine reservists and as many as 40 percent of Air Force reservists lived within 25 miles of their drill unit. In addition, the tables list the percentage of those residing at least 100 miles away from the drill unit or the closest installation. For service members, as many as 31 percent of them resided at least 100 miles away from their drill unit, and as many as 27 percent were at least 100 miles away from the nearest installation. In both cases, the Army Reserve had the highest proportion of service members 100 or more miles away, while the Air Force Reserve had the lowest proportion. Among the spouses, the relative ranking of the components was similar, with a greater proportion of Army Reserve spouses 100 or more miles away and fewer Air Force Reserve spouses in similar circumstances, but the actual percentages differed from the service member portion of the sample. For example, only 2 percent of Air Force Reserve spouses were 100 or more miles away from the nearest installation, compared with 12 percent of the Air Force reservists in our study.

The final items included in Tables 2.3 and 2.4 pertain to the most recent OCONUS deployment experienced by the service members or spouses. Since family readiness is likely influenced by the amount of notice the family receives, we obtained this information during our interviews. It is reflected in two measures summarized in the tables: the percentage of service members (or spouses) who received one week's notice or less and the percentage of service members (or spouses) who received one month's notice or less. As shown in Table 2.3, about one-fifth of the Marine reservists and Air Force reservists in our study indicated they received one week's notice or less, and the majority of reservists in three of the four components we studied reported receiving one month's notice or less. Army guardsmen fared the best in this respect: Only 6 percent of them received one week's notice or less, and 42 percent of them received one month's notice or less. Turning our attention to the spouses in our study, the situation is somewhat better: Compared with the interviewed service members, the percentages of families receiving one week's notice or less is smaller for all four components. While a sizable minority of spouses from all four components reported receiving one month's notice or less (ranging from 39 percent for the Marine Forces Reserve to 49 percent for the Army Reserve) their proportions were smaller than for the service members in our sample. Even so, Army reservists and spouses were the most likely to report receiving one month's notice or less.

The last two measures listed in Tables 2.3 and 2.4 pertain to the length of the most recent deployment. For both the service member and spouse portions of the interview sample, the vast majority of those from the Army National Guard and Army Reserve indicated that their most recent deployment was one year or longer. Much smaller proportions of those from the Marine Forces Reserve and Air Force Reserve experienced deployments of that length. The difference between the two groups of reserve components is perhaps better illustrated with the second measure of deployment length, the average length. For the Marine Forces Reserve (both spouses and service members), the average length of the most recent deployment is eight months. The comparable value for the Air Force Reserve is four months. Figures for the Army National Guard and Army Reserve are again much higher, with the Army National Guard experiencing the longest deployments on average. These figures are consistent with remarks made by the military family experts we interviewed. They discussed the variation in deployment length between the reserve components and referred to the implications this could have for families, as shown in the comments that follow:

> The Army National Guard needs a little more help only because their deployments last longer—could be one year or more. (6: DoD military family expert)

> I do believe that the Air Force [Reserve] has a shorter length of deployment, which might make things different for the families. (15: DoD military family expert)

> The Army is generally 18 months away from the family, but then they have less frequent call ups. The Marines combine their training and deployment into a one year period (seven months deployed, three months trained), and this means less time away but more frequent call ups. . . . The Air Force, they are mobilized as units and they have shorter rotation periods. This must reduce the level of stress. (23: Non-DoD military family expert)

These remarks suggest that the deployment lengths of those in our interview sample may be similar to those for the overall reserve components from which the sample was drawn.

Discussion

In this chapter, we delved more deeply into comparisons between the Active Component and the Reserve Component and among each of the reserve components. All in all, a review of attributes related to gender, age, and children suggests that not only do the reserve component families look different from active component ones, but also that there are differences *among* the reserve components. The Army Reserve and Air

Force Reserve have a notable proportion of female reservists, while the Marine Forces Reserve has a larger share of young families. The results of our expert interviews indicate this may in turn influence how reserve component families experience deployments and what types of family support they need.

With respect to our interview sample, data indicate that the spouses and service members we interviewed tended to be older, on average, than the spouses and service members in the same reserve components. However, figures for the average number of children under 18 were often higher within the sample than for the overall components. With respect to gender, there were proportionately fewer female service members and male spouses in our study than are present in the reserve components in general. These differences may have implications for the results of our study; findings related to parenting and children may be stronger within our sample, for instance, while those related to gender may be less apparent.

Additional information about our interview participants provides insights about their maturity, the strength of their marriage, and their experience. Marriage length, particularly the proportion of newlyweds in our sample, may serve as a proxy for both maturity and relationship strength, while prior military experience and repeat deployments since 9/11 are important indicators of guard and reserve families' familiarity with the military in general and deployments in particular. Measures pertaining to employment status and family finances revealed that most of the spouses and service members we interviewed were employed and financially comfortable, and measures of distance highlighted the relatively large proportion of reserve families living at least 100 miles away from the drill unit or closest military installation. Finally, items related to the length of the most recent OCONUS deployment revealed that interview participants affiliated with the Army Guard and Army Reserve experienced considerably longer deployments, on average, than did their counterparts in the Marine Forces Reserve and Air Force Reserve.

Overall, the variation within our interview sample along many of these dimensions, coupled with their apparent importance as suggested by the military family experts we interviewed, indicated that these were important attributes to focus on in our analysis. Throughout the remainder of the report, we discuss the extent to which statistically significant patterns were present in these data on individual and situational characteristics.

CHAPTER THREE

How Ready Are Guard and Reserve Families?

Family readiness is regarded as a critical aspect of preparedness for a service member's active duty service. DoD has stated that "The Department's ability to assist service members and their families to prepare for separations during short and long term deployments is paramount to sustaining mission capabilities and mission readiness" (Office of the Assistant Secretary of Defense for Reserve Affairs [RA], no date). Accordingly, family readiness was extensively addressed in the *National Guard and Reserve Family Readiness Strategic Plan for 2000–2005* (RA, 2002) and has been regularly assessed at various levels, including in such large-scale surveys as DMDC's 2006 Survey of Reserve Component Spouses and RA's 2002 Survey of Spouses of Activated National Guard and Reserve Component Members.

However, how family readiness is defined and measured varies. For instance, DoD Instruction 1342.23 suggests that family readiness encompasses a range of issues and tasks, including but not limited to wills, power of attorney, and other legal matters; financial management issues; family care plans; information about the unit's mission and expected activation; employment or reemployment rights; and "predictable psychological strains associated with military service" (DoD, 1994, p. 4), yet these specific aspects of readiness are not generally assessed. In the aforementioned surveys, one overarching question is posed to the spouse about the family's level of readiness or preparedness. Additionally, recent Status of Forces surveys of reserve component service members tend to overlook this subject entirely.[1] While this may be due to the need to keep surveys at a reasonable length and to avoid respondent fatigue, it is not clear whether a single measure of family readiness can be relied on for this important construct. As one of the participants in our military family expert interviews explained,

> It is really hard to get a quantitative measure of what family readiness means. The basic question that needs answering is, Can the family function while the service member is gone and do they have the tools and network necessary to get assistance? Do they have a power of attorney, are they enrolled in benefits, how do they stack up in terms of the standard administrative checklist? The more that these aspects are in order, the more ready the family is. (5: DoD military family expert)

Moreover, when readiness checklists or lists of tasks completed are provided in surveys (e.g., May 2003 Status of Forces Survey of Reserve Component Members, 2006 Survey of Reserve Component Spouses) they do not include measures pertaining to the psychological or mental health aspects of preparedness to which the aforementioned DoD Instruction refers. Rather, their emphasis tends to be on financial or legal elements of readiness.

Given the qualitative nature of our study, we had the opportunity to explore what family readiness means, not only to spouses, but also to the service members themselves. The potential link between family readiness and mission or unit readiness suggests that a service member's perception of his or her family's level of readiness may be an important metric to consider even though the service member is geographically separated from the family during a deployment. Specifi-

---

[1]   Status of Forces Surveys of Reserve Component Members from May 2003, September 2003, May 2004, November 2004, and June 2005 were reviewed to make this determination. Survey instruments were examined, with special attention paid to questions using the terms "readiness," "ready," "preparation," "preparedness," or "family." In none of these surveys was the *level* of family readiness or preparedness directly measured. The May 2003 survey, which included a list of questions related to family preparedness, was the only service member survey reviewed that addressed this topic, yet even that instrument did not include a measure of overall family readiness.

cally, we asked both spouses and service members, "What does it mean for your family to be ready for activation or deployment?" We posed the question without prompts; no examples of readiness were provided. For some, this was a bit of a challenge, something they seemed unclear about or wanted more time to consider, but ultimately, 79 percent of the spouses and service members in our study defined what family readiness meant for their family. Also of note, the proportion of service members who described family readiness was similar to the proportion of spouses that did so, suggesting that service members knew enough about their family's efforts to prepare for their deployment to offer a definition for family readiness.

Defining Family Readiness

A variety of family readiness aspects were discussed by the spouses and service members in our study, and in a number of instances an individual mentioned more than one basis of family readiness. Overall, three types or components of family readiness were each cited by approximately two-fifths of interviewees: financial readiness, readiness related to household responsibilities, and emotional or mental readiness. When only those who offered a family readiness definition are considered, this figure is closer to one-half of interviewees. The proportion of interviewees who provided each of these three definitions of family readiness (among those who provided any definition) is shown in Figure 3.1.

Legal preparedness was also mentioned by 25 percent of spouses and service members overall, and by 32 percent of those who defined family readiness in any way. Military resource–related readiness, which pertained to learning about and accessing military programs and benefits, such as Family Readiness Groups and TRICARE, was mentioned by 10 percent of the study sample, or 13 percent of those who provided a family readiness definition. A similar proportion of participants also discussed readiness in terms of getting a support system in place, so that the family would know whom to go to for necessary information (both military and nonmilitary) as well as for emotional support. In

Figure 3.1
Definitions of Readiness Provided by Service Members and Spouses

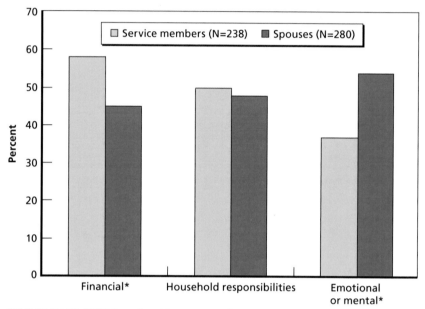

SOURCE: RAND 2006 Guard and Reserve Family Interviews.
*The service member and spouse percentages for this item were statistically different from one another at p<0.10.
RAND MG645-3.1

the following paragraphs, we provide exemplary comments for each of these dimensions of family readiness and discuss the characteristics of those who tended to mention a particular type of readiness. Additional aspects of readiness mentioned by less than 10 percent of the overall interview sample include employment-related arrangements, such as tending to one's own business or notifying an employer; finding out information regarding the deployment; determining ways for the service member and the family to communicate during the deployment; canceling classes, canceling vacation plans; and preparing the service member for the deployment.

Financial Readiness

Financial readiness includes an assortment of financial tasks, including saving money in anticipation of a break in pay (in between civilian and military paychecks) or in case of emergency, notifying creditors, and both short and long-term financial planning. Typical comments include the following:

> We had enough time to arrange financially for me to go on active duty. There was a break in money coming into the house until I started getting paid, and we were able to plan well enough to where it did not adversely affect us. (282: Army National Guard, O-3)[2]

> Making arrangements with banks and creditors. (482: Army National Guard, E-4)

> It means that we have paperwork in line, and we have our finances figured out. We know exactly what our pay would be. We know exactly what our BAH [Basic Allowance for Housing] would be, what our COLA [Cost of Living Adjustment] would be, what all of our allotments would be so that we could budget accordingly. (525: Army Reserve, E-5's wife)

> To be financially prepared in case we have to take a hit for any like major repair or anything while he is away. (647: Marine Forces Reserve, E-6's wife)

For service members, this was the most frequently cited type of family readiness, mentioned by 58 percent of the service members who offered a definition of family readiness when prompted during their interview. This was a statistically significant contrast to the spouses we

---

[2] Similar to the expert interviews, after each quotation from a spouse or service member interview, a unique identifier indicates the interview in which the comment was made. The same identifier is used to denote the same interview throughout the report, but it does not have significance nor can it be used to identify the participants. These numerical identifiers are used to convey the extent to which evidence is present in multiple interviews. We also include relevant demographic information; for example, we note whether the service member is female.

interviewed; 45 percent of spouses who defined family readiness did so in terms of financial preparation, and it was their third most frequently cited type of readiness. This is shown in Table 3.1, which also lists the characteristics along which service members, spouses, or both groups significantly differed in how frequently they provided a financially oriented definition of family readiness. For example, in the service member portion of the sample, 67 percent of junior officers who defined family readiness mentioned its financial aspects, compared with 54 percent of mid-grade enlisted personnel. Within the service member portion of our sample, there was also a greater tendency among Army guardsmen and Army reservists to discuss financial aspects of family readiness when compared with Air Force reservists. Distance from the nearest military installation and repeat OCONUS deployments were also the basis for statistically significant differences among service members who defined family readiness in terms of finances. Specifically, those living farther away from the nearest military installation were more likely to do so than those living closer, and service members who reported only one OCONUS deployment tended to offer financially oriented definitions of family readiness more than did those who experienced multiple deployments.

Two of the factors on which differences in responses were based for service members—pay grade and reserve component—also suggest explanations for response patterns among the spouses who defined family readiness financially in their interview. Specifically, spouses married to junior officers were more likely to discuss financial aspects of family readiness than were those married to enlisted personnel. In addition, spouses married to personnel serving in either the Army National Guard or Army Reserve were more likely to offer this definition of family readiness than were spouses married to Air Force reservists. Service member prior active duty experience, distance from the drill unit, and use of military-sponsored support programs also help to account for differences in how spouses defined family readiness. Spouses of service members who lacked prior active duty experience were more likely to discuss financial aspects of family readiness. Spouses living farther away from the drill unit also mentioned financially oriented family readiness more often than those who were

Table 3.1
Characteristics Associated with a Financial Definition of Family Readiness

	Service Members (%)	Spouses (%)
Overall percentage providing definition (N=238 service members; N=280 spouses)	58	45
Service member pay grade		
E-1 to E-4 (N=52 service members; N=72 spouses)	56	43
E-5 to E-6 (N=114 service members; N=138 spouses)	54	39
O-1 to O-3 (N=72 service members; N=70 spouses)	67	57
Service member reserve component		
Army National Guard (N=76 service members; N=81 spouses)	62	49
Army Reserve (N=60 service members; N=70 spouses)	65	50
Air Force Reserve (N=60 service members; N=67 spouses)	45	33
Marine Forces Reserve (N=58 service members; N=62 spouses)	59	45
Service member prior active duty		
Yes (N=151)		40
No (N=129)		50
Distance from drill unit		
Less than 25 miles (N=68)		35
25 or more miles (N=182)		48
Distance from nearest military installation		
Less than 25 miles (N=93)	47	
25 or more miles (N=145)	66	
Use of military-sponsored support programs		
Used military programs (N=179)		50
Did not use military programs (N=101)		36

Table 3.1—Continued

	Service Members (%)	Spouses (%)
Repeat OCONUS deployments since 9/11		
No (N=200)	62	
Yes (N=38)	42	

SOURCE: 2006 RAND Guard and Reserve Family Interviews.

NOTES: Ns are provided for either service member or spouse, as denoted in the table. Ns represent the total number of those who provided a definition for family readiness. When data from both groups are shown, Ns are specified as service member or spouse. All percentages shown are statistically different from one another at p<0.10. Shading indicates a subset of population that is not significantly different from other subsets. For the reserve component comparisons, the Army National Guard and Army Reserve percentages were both significantly different from those of the Air Force Reserve. Other reserve component comparisons are not significantly different.

closer to the drill unit. Lastly, husbands and wives who reported that they used military-sponsored support programs[3] during their service member's most recent activation also cited aspects of financial readiness more frequently than did those who claimed not to have used such programs. They may have attended programs that addressed financial literacy, highlighted the importance of financial readiness, or otherwise helped them to prepare financially.

Household Responsibility–Related Readiness

This aspect of family readiness includes preparing to handle household responsibilities normally taken care of by the service member, be it bill paying, yard maintenance, or other chores, as well as making arrangements related to children, such as family care plans, child care, and everyday issues, such as who will pick up the children from school and

---

[3]   During the interviews, both spouses and service members were asked if they [their family] was aware of military-sponsored programs and services, and those who answered affirmatively were then asked if they [their family] used any such programs or services. Ninety-four percent of spouses indicated they were aware of military-sponsored programs and services, and 65 percent of spouses who were aware of these programs and services subsequently stated that they used at least one of them.

drive them to various extracurricular activities. The following comments illustrate this type of readiness:

> I think for us, the biggest issue was about being able to be prepared for daily household management kind of stuff. An example would be, do I know how to change the furnace filter, do I know how to check oil on vehicles, that kind of maintaining things while he's gone. That was kind of the biggest thing that really put us in a pinch, was making sure everything was prepared for winter, and that kind of stuff. (491: Army National Guard, E-5's wife)

> We had to find babysitters because I work full time and he doesn't. (207: Air Force Reserve, E-5's wife with two children)

> I had a lot of household or family duties that I take care of on my side of things that I had to get my wife started on and get her understanding who and what and where and why and so forth, and that was fairly time-consuming. (292: Army National Guard, O-3)

> To have a system in place to deal with all the changes, to deal with my absence, and all the things that are going to pop up while I'm gone. [This includes] a set of procedures, more or less, to take care of all the household bills, to take care of my bills, to make sure that the cars get serviced, to make sure that the grass gets cut, make sure that the exterminator is still scheduled. I guess just to maintain a status quo, I guess, a system to keep all those daily functions going. (262: Marine Forces Reserve, O-2)

Readiness related to household responsibilities was mentioned by similar proportions of both spouses and service members—48 percent of spouses and 50 percent of service members who defined family readiness mentioned this type of preparedness. As shown in Table 3.2, in both groups, there appeared to be pay grade–related differences in who mentioned household issues as they defined family readiness. Among the service members, junior enlisted personnel were less likely than mid-grade enlisted personnel to mention this definition, while for

Table 3.2
Characteristics Associated with a Household Responsibility–Related Definition of Family Readiness

	Service Members (%)	Spouses (%)
Age		
25 or less (N=42)		36
26 or more (N=238)		50
Gender		
Male (N=9)		89
Female (N=271)		47
Marriage length		
2 years or less (N=54)		28
3 years or more (N=226)		53
Service member pay grade		
E-1 to E-4 (N=52 service members; N=72 spouses)	37	42
E-5 to E-6 (N=114 service members; N=138 spouses)	57	46
O-1 to O-3 (N=72 service members; N=70 spouses)	50	60
Service member reserve component		
Army National Guard (N=81)		51
Army Reserve (N=70)		43
Air Force Reserve (N=67)		63
Marine Forces Reserve (N=62)		36
Distance from nearest military installation		
Less than 25 miles (N=122)		55
25 or more miles (N=135)		42

SOURCE: 2006 RAND Guard and Reserve Family Interviews.

NOTES: Ns are provided for either service member or spouse, as denoted in the table. Ns represent the total number of those who provided a definition for family readiness. When data from both groups are shown, Ns are specified as service member or spouse. All percentages shown are statistically different from one another at $p<0.10$. Shading indicates a subset of population that is not significantly different from other subsets. For the pay grade comparisons in the spouse group, the E-1 to E-4 and E-5 to E-6 categories are both significantly different from the O-1 to O-3 category. The other pay grade comparison is not significantly different. For reserve component comparisons in the spouse group, the Air Force Reserve is significantly different from those of the Army Reserve and the Marine Forces Reserve, and the Marine Forces Reserve is also significantly different from the Army National Guard. The other reserve component comparisons are not significantly different.

spouses, those married to either junior enlisted or mid-grade enlisted spouses cited household-related readiness more frequently than spouses married to junior officers.

No other statistically significant patterns were evident in the service member portion of the sample, but other patterns emerged among the spouses who defined family readiness. For instance, younger spouses (age 25 or less) and newlywed spouses (married 2 years or less) were less inclined to mention household responsibilities than either older spouses or those in longer marriages, respectively. It is possible they have fewer household issues to address before a deployment than older service members and spouses, who may have elderly parents in need of care, more or older children, and home-related responsibilities stemming from a larger residence or home ownership. Interestingly, husbands were also more inclined to mention household-related readiness than were the wives. Although the number of husbands in our sample was small, this gender-based difference was statistically significant. Reserve component and distance measures were also associated with spouses' tendency to discuss household-related responsibilities within the context of defining family readiness. To elaborate, Air Force Reserve spouses were more likely than either Army Reserve or Marine Forces Reserve spouses to offer this type of family readiness definition, and Army National Guard spouses tended to do so more often than Marine Forces Reserve spouses. With respect to distance, spouses living closer to the nearest military installation were more likely to discuss household responsibility–related family readiness than were spouses living farther away.

Emotional or Mental Readiness

Comments pertaining to emotional or mental readiness include a number of references to "being mentally ready" or having enough time for all family members, including children, to "deal with" the fact that the service member will be separated from his or her family for potentially a considerable length of time. The remarks that follow are consistent with this theme:

To be mentally prepared to live without your spouse. To prepare your children that their father is going to be gone. To deal with all of the other practical and emotional issues associated with that kind of a deployment as best as you can, including the concern you have over your spouse being in a war zone. (254: Army Reserve, O-3's wife)

I think it [readiness] means a lot because if you're not ready, then you get caught off guard and you get upset, and you get stressed, and then if you're ready, then you just have the right emotional state to be able to have him leave and not have a nervous breakdown. (402: Army National Guard, E-4's wife)

Emotional preparation—just knowing that I was going to be gone for so long. [So] all members of the family—me, my wife, and my child—were able to be emotionally prepared to miss significant events like Thanksgiving and Christmas, which I did miss. (391: Air Force Reserve, O-3 with one child)

[The advance notice] gave us time, my wife and I, to sit down and have some time off together and actually start getting prepared, start talking about the deployment and preparing ourselves mentally for it. (4: Army National Guard, E-5)

As Table 3.3 shows, the spouses and service members in our study who defined family readiness differed significantly in their emphasis of this aspect. For spouses, emotional or mental family readiness was the most frequently offered definition of family readiness, mentioned by 54 percent of those who defined family readiness, while it was discussed by 37 percent of service members, and placed third behind financial and household matters in terms of frequency of mention. The relative importance of emotional readiness to spouses, as suggested by this higher frequency, implies that its absence from the family readiness checklist-type questions spouses are asked (e.g., those featured in the 2002 Survey of Spouses of Activated National Guard and Reserve Component Members) may hinder efforts to fully understand and measure family readiness.

Table 3.3
Characteristics Associated with an Emotional or Mental Health Definition of Family Readiness

	Service Members (%)	Spouses (%)
Overall percentage providing definition (N=238 service members; N=280 spouses)	37	54
Age		
25 or less (N=31)	61	
26 or more (N=206)	33	
Gender		
Male (N=9)		22
Female (N=271)		55
Marriage length		
2 years or less (N=40 service members; N=54 spouses)	53	65
3 years or more (N=180 service members; N=226 spouses)	32	51
Service member pay grade		
E-1 to E-4 (N=52)	50	
E-5 to E-6 (N=114)	27	
O-1 to O-3 (N=72)	43	
Service member reserve component		
Army National Guard (N=81)		47
Army Reserve (N=70)		54
Air Force Reserve (N=67)		55
Marine Forces Reserve (N=62)		61
Spouse prior military (spouses only)		
Yes (N=29)		28
No (N=251)		57
Distance from drill unit		
Less than 100 miles (N=172)	47	
100 or more miles (N=66)	33	

Table 3.3—Continued

	Service Members (%)	Spouses (%)
Distance from nearest military installation		
Less than 100 miles (N=199)	56	
100 or more miles (N=39)	33	

SOURCE: 2006 RAND Guard and Reserve Family Interviews.

NOTES: Ns are provided for either service member or spouse, as denoted in the table. Ns represent the total number of those who provided a definition for family readiness. When data from both groups are shown, Ns are specified as service member or spouse. All percentages shown are statistically different from one another at p<0.10. Shading indicates a subset of population that is not significantly different from other subsets. For service members, the E-5 to E-6 pay grade category was significantly different from the E-1 to E-4 and O-1 to O-3 categories. The other pay grade comparison is not significantly different.

Table 3.3 also reveals how spouses and service members tended to differ in terms of what characteristics were associated with an emphasis on emotional preparedness. For the service members in our study, there were significant differences in how frequently this definition was offered based on age, marriage length, service member pay grade, distance from drill unit, and distance from the nearest military installation. Specifically, younger service members, newlyweds, and those living within 100 miles of either their drill unit or the nearest military installation were more likely to define family readiness in terms of emotional or mental health than were service members who were older, married longer, or lived farther away, respectively. It is possible that the findings on age and marriage length are related to maturity. In a similar vein, both the junior enlisted service members and junior officers in our sample were more likely to discuss this type of readiness than were mid-grade enlisted personnel; only 27 percent of mid-grade enlisted service members focused on emotional preparedness, whereas 43 percent of junior officers and 50 percent of junior enlisted personnel who defined family readiness opted to describe it in these terms.

Turning our attention to the spouses in our sample who offered definitions of family readiness, the tendency to describe family readiness in terms of emotional or mental health differed by gender, reserve component, and spouse prior military experience. Wives were far more likely than husbands to do so: 55 percent of wives who defined family readiness highlighted emotional preparedness, compared with just 22 percent of husbands. As in the case of household responsibility–related readiness, although the number of husbands in our spouse sample was small, there was a statistically significant difference based on gender. Spouses from two reserve components were significantly different from one another in this respect as well. While 61 percent of Marine Forces Reserve spouses who discussed family readiness mentioned its emotional aspects, less than half—47 percent—of Army National Guard spouses did so. Statistical analysis also revealed a pattern based on spouse prior military experience: As one might expect, spouses with prior military experience were much less likely than those without such exposure to discuss family readiness as an emotional or mental health issue. Finally, just as newlywed service members were more likely to offer a family readiness definition that included emotional preparedness than were service members with longer marriages, so too were newlywed spouses.

Additional Definitions of Family Readiness

Legal preparedness, which refers primarily to drawing up a will and arranging for power of attorney, was mentioned by 29 percent of the spouses and 35 percent of the service members who provided a definition of family readiness. Typical comments include the following:

> It's the mundane stuff, the power of attorney, the will updated, and that sort of thing. (31: Army National Guard, O-3)

> [Readiness means] [a]ll legal paperwork to be taken care of so I can make all the legal decisions for the family. . . . That's the major thing, the legal ramifications of his absence. (481: Air Force Reserve, O-2's wife)

Spouse and service members also defined family readiness in terms of familiarizing themselves with military programs and services, especially TRICARE, with similar frequency. For example, as one spouse explained, "Basically [readiness is] just knowing what we needed to do as far as medical and things like that, about our benefits as far as the insurance and things like that go" (297: Marine Forces Reserve, E-5's wife). Spouses and service members differed significantly, however, in how often they discussed getting a network of support in place as a form of family readiness. Ten percent of spouses who defined family readiness provided this specific definition, compared with just 4 percent of service members. Representative comments from spouses follow:

> I think there should be a support system in place. That's the biggest thing, having a support system. (409: Army Reserve, O-3's wife)

> It means that we know the best way to contact him in case of an emergency, we know who to call and we're aware of the support system. It means that we have a support system and we make contact with other people in the unit. (266: Army Reserve, E-5's wife)

Since spouses are the ones who remain behind during a deployment, being aware of whom to turn to for information, for help, and for social support during that time was likely more salient to them. The individuals who participated in our expert interviews articulated this less-frequently mentioned but seemingly important aspect of family readiness quite well, as the following excerpt demonstrates:

> The most important aspect of readiness is that, at a minimum, the families have a name and a phone number to call when they need help. . . . For example, a simple flat tire can escalate to a spouse in emotional distress when a spouse just does not have the support to deal with issues. So, we decided to formalize the informal network of support that existed. We found that families need people to talk to; they need to know where they can access resources. (2: DoD military family expert)

Since smaller numbers of spouses and service members mentioned these dimensions of family readiness, fewer discernable patterns were present.

Readiness Levels of the Families in Our Study

Family Readiness Categories

After asking service members and spouses how they defined family readiness, we then used a second open-ended question to gauge how ready they felt their family was for their most recent deployment experience. While we could not independently verify these assertions, we contend that a family's perception of its readiness may be as important or even more important than an assessment based on objective criteria. During our coding and analysis, we organized responses to this question into three categories. The first category, "ready or very ready," included frequently short responses about being well prepared, without mention of anything for which the family was not prepared, such as statements about being "100% ready" or "pretty ready." The second category, "somewhat ready," consisted of comments about how the family was moderately ready except for an issue or two. It includes such assessments as the following:

> Well I'd say we were pretty ready other than the paper work. (84: Air Force Reserve, E-6's wife)

> On a 1-10 scale about a 7; we were expecting it, but not totally ready. (63: Marine Forces Reserve, E-4)

Last, the third category, "not at all ready," comprised remarks that simply declared they were "not ready at all" or more detailed descriptions, such as

> Not very ready. He had never been gone before, and we were only married for 10 months, and we weren't ready for him to go. (425: Army Reserve, E-5's wife)

> We weren't ready at all. It was just something that I never expected, me being in the National Guard, I just thought we would never be deployed outside of the United States. My wife has never been without me. (445: Army National Guard, E-6)

Overall, 65 percent of the service members and 60 percent of the spouses in our study indicated that their family was ready or very ready for their most recent deployment. Similar proportions of those were interviewed characterized their family as somewhat ready or not at all ready; each category includes approximately one-sixth to one-fifth of the service members and spouses interviewed. A comparison of service member and spouse responses separately revealed similar proportions for each of the three categories. These proportions, as well as the share of interview participants who did not provide an answer, are shown in Figure 3.2.

Since service members have not been asked in recent surveys to evaluate their family's level of readiness, we did not have a basis of comparison for their assessment. For spouses, the results of the 2002 Survey of Spouses of Activated National Guard and Reserve Component Members could have potentially served as a referent for the spouse

Figure 3.2
Family Readiness Levels, as Reported by Service Members and Spouses

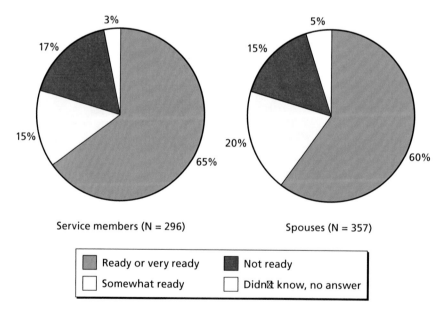

Service members (N = 296)   Spouses (N = 357)

| ■ Ready or very ready | ■ Not ready |
| □ Somewhat ready | □ Didn't know, no answer |

SOURCE: 2006 RAND Guard and Reserve Family Interviews.
RAND MG645-3.2

responses, but the question is worded somewhat differently from the one included in our interview protocol. While the 2002 survey gauged the spouse's level of preparedness upon first learning about a service member's call to active duty, our question pertained to the entire family and did not focus on a specific point of time. The emphasis on readiness at the time of notice may help to explain the smaller proportion of spouses who reported they were well prepared in the 2002 survey: 37 percent in 2002 compared with 60 percent of spouses in this effort. However, it is unclear whether or how the emphasis on the spouse's readiness instead of the whole family's readiness accounts for this difference. Spouses asked about their own level of preparedness, for instance, may not have considered their children's level of readiness in such an assessment.

Factors Related to Family Readiness Levels

The spouses and service member groups in our study resembled one another in their overall evaluations of family readiness levels but differed in terms of what factors were associated with family readiness. This is similar to family readiness *definitions*; patterns related to the type of definition offered tended to differ among the spouses and service members (i.e., characteristics that accounted for differences among spouses usually were not statistically significant among service members, and vice versa). Table 3.4 displays a subset of the characteristics that were associated with family readiness levels in our study. The relationship between activation notice and family readiness is summarized in the next section.

Two characteristics accounted for significant differences in the service member sample: age and marriage length, our proxies for maturity and marriage strength. Specifically, both newlyweds (those married for two years or less) and young service members (age 25 or less) less frequently characterized their family as ready or very ready and more frequently described them as not ready than did those married longer and older service members, respectively. Among spouse interviewees, patterns were apparent based on college degree, service member pay grade, and service member prior active duty status. Spouses lacking a college degree, those married to junior enlisted personnel, and those

Table 3.4
Characteristics Associated with Family Readiness Levels

	Service Members (%)			Spouses (%)		
	Ready or Very Ready	Somewhat Ready	Not at All Ready	Ready or Very Ready	Somewhat Ready	Not at All Ready
Age						
25 or less (N=37)	49	22	30			
26 or more (N=248)	70	14	16			
Marriage length						
2 years or less (N=49)	53	25	22			
3 years or more (N=215)	72	13	15			
College degree						
Yes (N=186)				70	19	10
No (N=153)				54	22	24
Service member pay grade						
E-1 to E-4 (N=87)				46	31	23
E-5 to E-6 (N=161)				70	17	13
O-1 to O-3 (N=91)				68	17	15
Service member reserve component						
Army National Guard (N=99 service members; N=97 spouses)	69	16	15	52	32	17
Army Reserve (N=72 service members; N=86 spouses)	56	17	28	61	17	22
Air Force Reserve (N=59 service members; N=81 spouses)	76	14	10	75	17	7

Table 3.4—Continued

	Service Members (%)			Spouses (%)		
	Ready or Very Ready	Somewhat Ready	Not at All Ready	Ready or Very Ready	Somewhat Ready	Not at All Ready
Marine Forces Reserve (N=56 service members; N=75 spouses)	70	13	18	68	13	19
Service member prior active duty						
Yes (N=197)				67	16	17
No (N=142)				58	27	16

SOURCE: 2006 RAND Guard and Reserve Family Interviews.

NOTES: Ns are provided for either service member or spouse, as denoted in the table. When data from both groups are shown, Ns are specified as service member or spouse. All percentages shown are statistically different from one another at p<0.10. Shading indicates a subset of population that is not significantly different from other subsets. For the pay grade comparisons in the spouse group, the E-5 to E-6 and O-1 to O-3 categories were both significantly different from the E-1 to E-4 category. The other pay grade comparison is not significantly different. For reserve component comparisons in the spouse group, the Army National Guard is significantly different from the other three reserve components. In addition, the Air Force Reserve is significantly different from the Army Reserve. The other reserve component comparisons are not significantly different.

married to service members without prior active duty experience all had less favorable views of their family's readiness than their counterparts with more education, higher rank, or more experience. For instance, 46 percent of junior enlisted spouses reported their family as ready or very ready, compared with 70 percent of mid-grade enlisted spouses and 68 percent of junior officer spouses.

The one attribute by which family readiness levels differed significantly for both service members and spouses was reserve component. For the service member portion of the sample, there was a notable difference in family readiness reported by Army reservists and Air Force reservists; while 76 percent of Air Force reservists reported that their family was ready or very ready for deployment, the proportion of Army reservists expressing a similar sentiment was significantly lower—but still relatively high—at 56 percent. In addition, 28 percent of Army reservists stated that their family was not ready at all, compared with only 10 percent of Air Force reservists. Perhaps these results are due, at least in part, to the high proportion of Army reservists in our study who received one month's notice or less, 68 percent, compared with members of the other three components included in our study (refer to Table 2.3 for detail). Within the spouse group of interview participants, Army National Guard spouses were significantly different from the other three components in their assessment of family readiness. This is best reflected by comparing the proportion of spouses who described their family as ready or very ready: 52 percent of Army National Guard spouses felt this way, compared with 61 to 75 percent of spouses from other reserve components. Further, the Air Force Reserve spouses differed from Army Reserve spouses in this respect in that a greater proportion of Air Force Reserve spouses stated their family was ready or very ready and a lower proportion of Air Force Reserve spouses viewed their family as not ready at all.

The preceding discussion suggests that the amount of notice may have influenced the level of family readiness for Army reservists. This is consistent with additional analysis we conducted. We evaluated family readiness levels in terms of the amount of notice families actually received and their perceptions of notice adequacy. Both distinctions are important, because families receiving the same amount of notice may have different opinions about its sufficiency. Accordingly, during the interviews, we not only asked participants for a quantitative measure of how much notice their family received, but we also asked an open-ended question about whether the amount of notice had an effect on the family. Responses to the first question were briefly summarized in Chapter Two and are provided in greater detail for both service members and spouses in Figure 3.3. For example, 14 percent of service members received seven days' notice or less, compared with only 5 percent of the spouses in our sample. At the other extreme, 22 percent of spouses indicated they had more than three months' notice (91+ days), while only 13 percent of service members enjoyed that much notice.

In response to the open-ended question about how the amount of notice affected their family, 63 percent of all interviewees (similar proportions of service members and spouses) provided answers related to its adequacy—i.e., whether they perceived the notice as adequate or

Figure 3.3
Activation Notice, as Reported by Service Members and Spouses

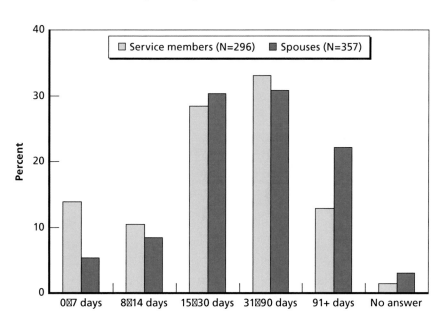

SOURCE: 2006 RAND Guard and Reserve Family Interviews.
RAND *MG645-3.3*

insufficient. Examples of responses from individuals who felt that they received adequate notice include the following:

> I think it was adequate. I think any more time to lull it over would have been bad, and I think one month was enough time that we could start looking at what we needed to do, and then while I was in training, handling any other minor problems that came up. It was always in the back of my head as a contingency, something to deal with, I wasn't surprised by it. It was something we had planned for prior to being notified officially. (118: Army National Guard, E-5 with one month's notice)

> We were able to get our affairs in order and everything was taken care of by the time he left. We didn't have to run around like crazy and get things signed. I think it also helps you prepare emotionally. Before his first call-up, we had three days and that

just doesn't give you enough time to process it emotionally, never mind running around. So this time we were more at peace with it by the time it came. (124: Marine Forces Reserve, E-5's wife with three months' notice)

It [the amount of notice] helped me get my belongings together, you know, around the house. It gave us time to go get the things that we needed, like military IDs or wills. Which we did have most of that but we had time to just double-check everything. It gave him time to give a notice at work. It gave me time to find child care. It was a good amount of time. I think anything less would have been rushing and anything more would have been difficult to wait. (315: Army Reserve, E-6's wife with two weeks' notice)

On the other hand, the following comments illustrate perceived insufficient notice:

Two weeks to go home, see my parents, tell my job I'm leaving, pay all my bills, transfer power of attorney, you name it. Two weeks notice is not enough time. The soldiers I had in my unit had 72 hours in some cases. Two weeks is not enough time. (574: Army Reserve, O-2 with two weeks' notice)

I was not happy about it [the amount of notice] at all. . . . When they called us, they talked about maybe he was going to be deployed, and then they said he just needed to come fill out paperwork. Instead of him just having to come fill out paperwork, he had to come fill out paperwork and stay down there. So we knew that he might, there was a chance, and then all of a sudden, it was the day before that he found out he was leaving. So it was kind of like a two-day notice. I didn't have anything in order. (174: Air Force Reserve, E-4's wife with two days' notice)

We didn't feel prepared. It kind of came at us from nowhere. We didn't have a whole lot of time. We were getting married around then, and it was just kind of a short notice. (804: Marine Forces Reserve, E-4's wife with six weeks' notice)

It affected me because they didn't give me enough time. They didn't give me enough time to secure everything that I needed to secure financially with my wife and medical and all that stuff. It just wasn't enough time. (105: Army National Guard, E-6 with one month's notice)

Table 3.5 summarizes the characteristics by which perceptions of notice adequacy differed significantly among those who discussed it. They include four attributes for both the service member and spouse samples: college degree, service member pay grade, service member reserve component, and amount of notice. For both service members and spouses, those with a college degree tended to regard the amount of notice received as adequate; comparable proportions were lower for those without a college degree. With respect to pay grade, both junior enlisted personnel and junior enlisted spouses were more likely to perceive inadequate notice, compared with junior officer service members and to mid-grade enlisted spouses. In addition, interviewees affiliated with the Army Reserve tended to characterize the amount of notice as insufficient. Within the service member sample, the proportion of Army reservists who felt their notice was insufficient was considerably higher than the proportions in the other three reserve components. Within the spouse sample, Army National Guard and Army Reserve spouses were significantly more inclined to describe their notice as insufficient than were Air Force Reserve and Marine Forces Reserve spouses. The last common finding confirmed a solid association between the actual amount of notice received and perceptions of notice adequacy: the greater the notice, the higher the proportion of interviewees who characterized that notice as adequate. Note, however, that this is not a *perfect* association; for example, 47 percent of service members who received one month or less notice still felt it was adequate.

Two additional characteristics accounted for significant patterns among the spouses who discussed their perceptions of notice adequacy: parental status and repeat OCONUS deployments. Not surprisingly, spouses who were not parents were more likely to feel that the notice they received was adequate; 83 percent of spouses without children at home expressed this sentiment, compared with 69 percent of spouses

Table 3.5
Characteristics Associated with Perception of Notice Adequacy

	Service Members (%)		Spouses (%)	
	Insufficient	Adequate	Insufficient	Adequate
Parental status				
Has children (N=170)			31	69
No children (N=53)			17	83
College degree				
Yes (N=85 service members; N=126 spouses)	22	78	23	77
No (N=103 service members; N=97 spouses)	35	53	34	66
Service member pay grade				
E-1 to E-4 (N=47 service members; N=61 spouses)	36	64	36	64
E-5 to E-6 (N=85 service members; N=103 spouses)	31	69	21	79
O-1 to O-3 (N=56 service members; N=59 spouses)	21	79	31	70
Service member reserve component				
Army National Guard (N=65 service members; N=65 spouses)	19	82	34	66
Army Reserve (N=45 service members; N=57 spouses)	53	47	40	60
Air Force Reserve (N=38 service members; N=48 spouses)	21	79	17	83
Marine Forces Reserve (N=40 service members; N=53 spouses)	28	73	17	83
Amount of notice				
One month or less (N=88 service members; N=89 spouses)	53	47	53	47
More than one month (N=100 service members; N=129 spouses)	8	92	12	88

Table 3.5—Continued

	Service Members (%)		Spouses (%)	
	Insufficient	Adequate	Insufficient	Adequate
Repeat OCONUS deployments since 9/11				
No (N=140)			33	67
Yes (N=83)			19	81

SOURCE: 2006 RAND Guard and Reserve Family Interviews.

NOTES: Ns are provided for either service member or spouse, as denoted in the table. Ns represent the total number of those who provided an opinion regarding notice adequacy. When data from both groups are shown, Ns are specified as service member or spouse. All percentages shown are statistically different from one another at p<0.10. Shading indicates a subset of population that is not significantly different from other subsets. For reserve component comparisons in the service member group, the Army Reserve is significantly different from those of the other three reserve components. For reserve component comparisons in the spouse group, the Army National Guard and Army Reserve are each significantly different from both the Air Force Reserve and the Marine Forces Reserve. The other reserve component comparisons are not significantly different.

with children. In addition, spouses married to service members who had more than one deployment since 9/11 were more likely to regard the notice received as adequate. This suggests that although deployment frequency was not directly related to family readiness, it potentially may influence family readiness *indirectly*, given its association with perceptions of notice adequacy.

As depicted in Table 3.6, our findings indicate that readiness levels were related not only to the actual amount of notice a family received but also to perceptions of its adequacy. For both service members and spouses, those receiving less notice were less inclined to describe their family as ready and more inclined to describe them as unprepared than those with more advance notice. To illustrate, about three-fourths of interviewees who received more than one month's notice reported that their family was ready or very ready, compared with roughly one-half of those with one month's notice or less. Further, about one-quarter of those with one month's notice or less described their family as not at all ready, compared with less than one-tenth of those with more than one month's notice. Similar findings were noted with respect to perceptions of notice adequacy: over four-fifths of interviewees who regarded

Table 3.6
Additional Characteristics Associated with Family Readiness Levels: Amount of Notice and Perception of Notice Adequacy

	Service Members (%)			Spouses (%)		
	Ready or Very Ready	Somewhat Ready	Not at All Ready	Ready or Very Ready	Somewhat Ready	Not at All Ready
Amount of notice						
One month or less (N=149 service members; N=151 spouses)	55	18	27	49	23	28
More than one month (N=137 service members; N=179 spouses)	80	12	8	75	17	7
Perception of notice adequacy						
Insufficient (N=52 service members; N=59 spouses)	21	25	54	17	31	53
Adequate (N=132 service members; N=159 spouses)	85	11	4	82	16	2

SOURCE: 2006 RAND Guard and Reserve Family Interviews.
NOTES: Ns are provided for either service member or spouse, as denoted in the table. When data from both groups are shown, Ns are specified as service member or spouse. All percentages shown are statistically different from one another at p<0.10.

the amount of notice as adequate indicated that their family was ready or very ready, compared with about one-fifth of those who felt it was insufficient, and just over half of those who regarded the amount of notice as insufficient claimed their family was not at all ready, in contrast with a small number of spouses and service members who deemed the notice adequate.

Military Preparedness

As noted at the outset of this chapter, part of the reason family readiness is viewed as important is its relationship with unit or mission readiness. Although we could not assess military readiness at those levels, we did ask the service members in our study about their own level of military preparedness prior to their deployment with a question adapted from recent Status of Forces Surveys: "Overall, how well prepared were you to perform your active duty job during your most recent activation?"[4] A five-point scale, ranging from very well prepared to very poorly prepared, was provided, which we collapsed to a three-point scale in our analysis for ease of presentation. Overall, 79 percent of service members indicated they were well prepared, 11 percent stated they were neither well nor poorly prepared, and 9 percent described themselves as poorly prepared. We noted a small number of significant differences based on demographics or other individual attributes, provided in Table 3.7. One indicator of maturity, marriage length, was related to military preparedness; newlyweds were less likely to say that they were well prepared than were those in longer marriages. In addition, both the actual amount of notice received and perceptions of its adequacy were significantly associated with service members' military preparedness. As one might expect, those who received more notice and those who felt it was adequate tended to characterize themselves as well or very well prepared for their active duty job at higher rates than did those who received less notice or perceived it as insufficient. Further, while the overall number of service members who described themselves as poorly or very poorly prepared was small, they were more likely to have less actual notice and/or perceptions that their notice was insufficient.

Lastly, there was a strong interrelationship between family readiness and military preparedness. Since our interview data are cross-sectional,

---

[4]   For example, the precise wording in the June 2005 Status of Forces of Reserve Component Members was, "Overall, how well prepared are *you* to perform your wartime job?" We used the past tense to reflect that the service members we interviewed had been demobilized, and we opted to focus the question on their active duty jobs rather than their wartime jobs, a potentially subtle distinction.

Table 3.7
Characteristics Associated with Military Preparedness

	Service Members (%)		
	Well or Very Well Prepared	Neither Well nor Poorly Prepared	Poorly or Very Poorly Prepared
Marriage length			
2 years or less (N=50)	66	22	12
3 years or more (N=221)	81	10	9
Amount of notice			
One month or less (N=155)	77	10	13
More than one month (N=140)	82	12	6
Perception of notice adequacy			
Adequate (N=133)	83	12	5
Insufficient (N=54)	59	19	22
Family readiness			
Ready or very ready (N=192)	88	9	4
Somewhat ready (N=42)	67	21	12
Not at all ready (N=51)	61	12	28

SOURCE: 2006 RAND Guard and Reserve Family Interviews.
NOTE: All percentages shown are statistically different from one another at $p<0.10$.

such that both measures were obtained at the same time and from the same person, we could not determine whether one type of readiness affected the other, or whether a third factor, such as an underlying personal attribute, influenced both family readiness and military preparedness. Eighty-eight percent of service members who regarded their family as ready or very ready for deployment also indicated that they were well or very well prepared for their active duty job, while only 61 percent of service members whose families were not ready at all graded their own military preparedness highly. Conversely, 28 percent of service members who said that their family was not ready also indicated that they were poorly or very poorly prepared from a military standpoint, whereas only 4 percent of service members who described their family as ready or very ready had a poor opinion of their own military preparedness. While we could not independently ascertain the military readiness of the service members in our study, these findings provide some evidence of the much-discussed link between family readiness and unit or mission readiness. Further, they support the notion that family readiness is not only important from the standpoint of the social compact, but also in terms of military effectiveness.

Discussion

Although family readiness is depicted in DoD instruction as a multi-faceted concept, few large-scale surveys have attempted to measure it as such. Recent surveys of reserve component spouses typically gauge family readiness using one item, while reserve component service members themselves are generally not asked in Status of Forces surveys to assess their family's level of readiness. Our qualitative research approach allowed us to explore how both spouses and service members define family readiness and to consider whether present efforts to measure family readiness were adequate given its perceived relationship with mission readiness. The results of our interviews revealed that both spouses and service members identified three main dimensions of family readiness: financial readiness, readiness related to household responsibilities, and emotional or mental readiness. However, service members mentioned financial aspects of family readiness more often than did spouses, while spouses in turn were more focused on emotional or mental readiness. Spouses not only discussed this type of preparedness more frequently than did service members, but it was also their most commonly offered definition of family readiness. The relative importance of emotional or mental readiness to spouses suggests that its absence from the family readiness checklist-type questions spouses are typically asked may hinder efforts to fully understand and measure family readiness. Moreover, the differences between spouses and service members indicate that relying on one of them to evaluate their family's level of readiness, as appears to be the current practice, may result in an incomplete or otherwise inaccurate assessment.

We also asked spouses and service members to characterize their family's level of readiness, and the majority of both indicated that their family was ready or very ready for the most recent deployment. Table 3.8 summarizes the factors associated with family readiness levels and shows whether each relationship was present in the service member portion of the sample, the spouse portion, both, or neither. Specifically, spouses with a college degree, spouses married to a service member with prior active duty experience, interviewees (both spouses and service members) who received more notice, and interviewees who perceived the amount of notice to be adequate were all more likely to characterize their family as ready or very ready, while spouses lacking a college degree, spouses married to a service member without prior active duty experience, those receiving less notice, and those who felt the notice was insufficient were more likely to claim that their family was not ready at all. Conversely, junior enlisted spouses, service members age 25 or younger, and newly wedded service members (married two years or less) tended to describe their family as not ready more frequently than spouses married to personnel in higher ranks (mid-grade enlisted and junior officer), service members age 26 and up, and service members married longer than two years. Both Air Force reservists and Air Force Reserve spouses were more likely than Army reservists and Army Reserve spouses to describe their family as ready or very ready. Further, spouses married to Army guardsmen were less inclined to characterize their family as ready or very ready than were spouses affiliated with the other three reserve components. Several potentially important factors we considered throughout our analysis—gender, parental status, spouse prior military experience, distance measures, and deployment frequency—were not statistically related to family readiness levels for either the service members or the spouses included in our study.

Lastly, we examined the relationship between family readiness and military preparedness for the service members we interviewed. Since surveys of reserve component personnel typically do not measure family readiness, we viewed this as an important opportunity to substantiate the link between family readiness and at least one type of military preparedness, that of the individual service member. Our analysis demonstrated that there was a strong interrelationship between

Table 3.8
Summary of Factors Related to Family Readiness Levels

	Portion of Interview Sample
Individual and situational characteristics	
Age	SM
Marriage length	SM
College degree	SP
Service member pay grade	SP
Service member reserve component	SM, SP
Service member prior active duty	SP
Amount of notice	SM, SP
Perception of notice adequacy	SM, SP

SOURCE: 2006 RAND Guard and Reserve Family Interviews.

NOTES: All relationships listed are statistically significant at p<0.10. SM = Finding present in the service member portion of the sample (N=296); SP = Finding present in the spouse portion of the sample (N=357).

family readiness and military preparedness: Service members who said they were well prepared for active duty were more likely to characterize their family as ready or very ready, and those who believed they were poorly prepared for active duty tended to feel that their family was not ready at all. Although we could not assess causality or verify military readiness from an external source, these results offer some support for the link between family readiness and mission readiness. They also suggest an additional benefit to collecting family readiness data from service members and not only spouses—the ability to substantiate this relationship further.

CHAPTER FOUR

What Problems Do Guard and Reserve Families Report?

This research effort explored the problems and challenges faced by reserve component families. When we asked experts on reserve family issues about problems that they believed these families confront, the majority of military family experts indicated that reserve component families experience the following problems: financial problems, health care issues, emotional or mental problems, and household responsibility issues. The expert discussions of problems faced by guard and reserve families included the comments that follow. More than the other problems, the experts tended to mention financial problems. These comments were typical:

> They have problems with the pay system or understanding the pay system. (7: DoD military family expert)

> Finances are also a problem. Some families go from an $80,000 salary to a $40,000 salary. Who is going to make up the difference? Sometimes an employer will, but this is not the norm. Financial problems increase stress and can cause many other problems for families. . . . [W]e all live on credit, and we all have car loans, student loans, and these do not go away when the service member changes jobs. There is a lot of emotional stress that comes with these financial difficulties. (2: DoD military family expert)

> Managing the change in finances is also difficult. Some families might actually be better off with their change in salaries, but

some have the alternative effect and are faced with dire financial issues. Trying to manage the change in pay, especially when a family is accustomed to having a certain standard of living, can create stress and other strains on the family. (1: DoD military family expert)

Many of the health care issues were related to the transition to and from the TRICARE system:

When a reservist is deployed, the military picks up their health insurance coverage and this continues for them for a certain amount of time. This continuation/discontinuation is a big problem for reservists. A horror story that I am hearing far too often is the one where the service member goes off to deployment, the family is forced to change insurance companies, and when they are sent back home, they are expected to pick up the prior coverage again. Well, say the wife has diabetes. Now the new insurance coverage considers her condition pre-existing, even though they covered it for 15 years prior to her stint on TRICARE. They treat her like a brand new enrollee. (16: non-DoD military family expert)

Another problem is the health care issues they deal with. Most employers don't continue care for families once the employee leaves for activation. This means that families have to change doctors, if doctors who take TRICARE are even available in their areas. This leaves families with confusion about the system, and creates a health care need that can include not getting services, not knowing where to get services, and owing more money than they would have otherwise. (2: DoD military family expert)

The emotional or mental issues mentioned included the strains and difficulties experienced by the spouse and the children as well as the returning service member:

First, the husband and wife may live in an area where they may not be connected to close family nearby, so [in the event of activation/deployment] the stay-behind spouse is in an isolated location with no support. They feel stranded, abandoned, alone; it affects their emotions. (6: DoD military family expert)

There are mental health issues to deal with. There is a lot of stress that is part of deployment on the families, separation when a spouse leaves, and the anxiety of a spouse being in a war zone. We are also seeing greater levels of PTSD [post-traumatic stress disorder] [among returning service members]. (5: DoD military family expert)

Kids have problems with deployment as they wonder whether they will be safe [given their service member parent is not home to protect them from those who may target the U.S.], whether things will be the same when the service member comes home, what to say to them while they're gone. (7: DoD military family expert)

Another one we hear about more frequently now affects children, especially school-age children in large-population schools that may not have experience with children of deployed military members. These children may be ostracized [as a result of their unique situation] or, even if they seek help, the counselors and teachers don't know how to help them. (3: DoD military family expert)

The household responsibility problems mentioned by the experts focused on the ability to maintain a household and continue the various children's activities, without the support and effort of both parents. These comments included the following:

The family may have had a system for things like picking up the kids from day care. That [system] no longer works. The wife may feel she has to quit her job to be with the kids and she will feel like a single parent. (4: DoD military family expert)

The needs that come from a spouse having to run a household alone. Needs can include having to find a way to manage the new constraints on the work schedule, get the kids to after school activities, dealing with sick children, managing the household, alone, when they were always working as a couple. Playing the parenting role is difficult alone. (1: DoD military family expert)

The process into which they become a military family is very quick and the first time a spouse has to figure out what to do when their car breaks down (get a tow truck, find out it takes $1,000 to fix) can be unnerving. It takes some time for the family to adjust to having a family member gone. But as the cycle progresses, they seem to adjust very well. As things become familiar, they become better able to handle these kinds of changes. . . . Upon return, we have [the problems associated with] reunion. . . . For example, the service member might have been the family member who paid the bills before, but for the year they are gone, the spouse takes over this job. The service member might expect to come back and take charge of the checkbook again, but that is not always the natural progression. (8: DoD military family expert)

We subsequently heard about many of these problems from service members and spouses themselves during our interviews. Our research provides insights into the nature of these problems, the extent to which they were experienced by our interviewees, and the characteristics of the families that were more likely to mention the different types of issues. Specifically, we asked spouses and service members, "What types of issues or problems did your family face, or is currently facing, as a result of [your or your spouse's] activation or deployment?" This question was initially posed without prompting in order to identify the problems most salient to the spouses and service members in our study and to avoid giving interviewees the opportunity to simply respond in the affirmative when presented with a list of issues. The majority of spouses and service members responded by mentioning some kind of problem stemming from deployment. After interviewees were given an opportunity to describe their issues and problems, the interviewer probed further by stating, "Issues that have been mentioned to us include those related to emotional stability, health care, employment for the service member or the spouse, education for the service member or the spouse, family finances, household responsibilities and chores, marital health, and children. Can you talk about the extent to which any of these have been issues with your family?" This analysis considers all issues provided by interviewees both before and after this interview probe. Figure 4.1 indicates the proportions of interviewees, broken down across spouses and service members, who mentioned different types of problems.

In all, 79 percent of interviewees mentioned a problem during the interviews. Spouses were more likely to mention emotional or mental problems, household responsibility problems, and children's issues than were service members. Thirty-nine percent of spouses mentioned emotional problems, approximately the same proportion mentioned household responsibility problems, and 26 percent of spouses mentioned children's issues. Ten to fifteen percent of spouses each mentioned financial/legal issues, employment issues, marital issues, and health care. Only about 5 percent of spouses mentioned education problems. Among service members, the most frequently mentioned prob-

Figure 4.1
Problems Reported by Service Members and Spouses

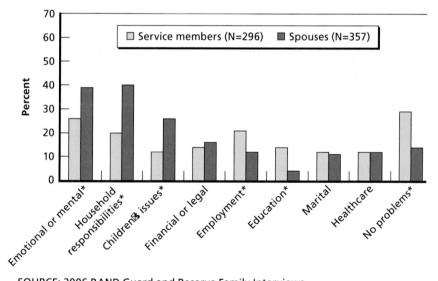

SOURCE: 2006 RAND Guard and Reserve Family Interviews.
*The service member and spouse percentages for this item were statistically different from one another at p<0.10.
RAND MG645-4.1

lems were emotional or mental problems (mentioned by 26 percent of service members), household responsibility issues, and employment problems (each mentioned by about 20 percent of service members). Financial and legal issues, education issues, marital problems, health care issues, and children's issues were each mentioned by roughly 10 percent of service members. The following sections feature exemplary comments from spouses and service members to help depict each problem and discuss the characteristics of those interview participants who tended to mention experiencing each kind of problem.

Emotional or Mental Problems

Emotional issues were mentioned by about one-third of all interviewees, and were mentioned significantly more frequently by spouses than by service members. When determining what kind of interview responses to include in this category, we purposely excluded comments that referred only to missing the reservist or guardsman. For example, we decided not to include a response such as "We're fine, we just missed him" as an indicator of emotional problems, based on the assumption that all families missed their service member. Nonetheless, the many answers that were included in this category suggest a range of severity, from relatively mild sadness and anxiety to more severe emotional or mental difficulties that required medical attention. In other words, the comments reflected a range of emotional difficulty, from "We would rather him be home. It's just emotionally inconvenient"(118: Marine Forces Reserve, E-3's wife), to "I went through a period of depression and I'm on an anti-depressant [medication]" (168: Army Reserve, O-3's wife) and "I've been on a lot of anti-depressants, and then he left, and he went over to Iraq, and I found out I was pregnant, and because of all the stress and depression, I've had another miscarriage, so I've had two" (316: Army Reserve, E-4's wife). Additional comments included

> Having to deal with the mental stress of not knowing what is going on with your spouse. Like with my wife, when she was pregnant she didn't know what was going on with me and she

would hear on the news that certain places got hit in Iraq, you know, and all that mentally will break you down, and that is just not a good thing for a pregnant woman. (283: Army Reserve, E-5)

[W]hile I was gone, my wife tried committing suicide 'cause she was just stressful and worried. (318: Army Reserve, E-5)

Unless a person has had to go through having a loved one deployed, they really can't understand what we go through. It's one thing to deal with death, but it's another thing to worry about if it's going to happen and when it's going to happen. It's like you can't live fully without having that constantly on your mind. It's a mental and emotional aspect that I wouldn't wish on anyone. (14: Army National Guard, E-4's wife)

[O]nce your husband is not at home, one day it's good, the other one it's better, the other day it's not so good. You have to stay strong for you and others around you, and face the people who are not so positive for this war. They don't support the soldiers like they have to or the Marines because they're against the war. So you know, you find yourself alone in this battle sort of, kind of. (58: Marine Forces Reserve, E-5's wife)

Table 4.1 includes the characteristics of interviewees that indicated emotional or mental problems stemming from deployment. Service members who reported more than one month of deployment notice and service members who stated that the notice they received had been adequate were less likely to mention emotional problems, as were spouses who said that their family had been ready for the deployment. Service members who said that their family was financially comfortable were also less likely to mention any emotional or mental problems.[1]

---

[1] Although the question pertaining to family finances asked about the family's *current* family situation, which was post-deployment for all of the service members and some of the spouses interviewed, we believed it was acceptable to consider in our analysis of the problems reported *during* deployment because the majority of both spouses (61 percent) and service members (72 percent) indicated that their family's finances had not changed as a result of their deployment.

Table 4.1
Characteristics Associated with Citing Emotional or Mental Problems

	Service Members (%)	Spouses (%)
Overall percentage citing emotional or mental problems (N=296 service members; N=357 spouses)	26	39
Age		
25 or less (N=55)		55
26 or more (N=302)		36
Marriage length		
2 years or less (N=71)		55
3 years or more (N=286)		35
Parental status		
Has children (N=269)		36
No children (N=88)		50
Service member reserve component		
Army National Guard (N=104 service members; N=102 spouses)	21	38
Army Reserve (N=74 service members; N=89 spouses)	35	46
Air Force Reserve (N=60 service members; N=83 spouses)	25	29
Marine Forces Reserve (N=58 service members; N=83 spouses)	26	43
Service member prior active duty		
Yes (N=206)		32
No (N=150)		49
Spouse prior military (spouses only)		
Yes (N=34)		21
No (N=323)		41
Financial situation		
Comfortable (N=195)	22	
Occasional difficulty (N=66)	35	

Table 4.1—Continued

	Service Members (%)	Spouses (%)
Uncomfortable (N=35)	34	
Amount of notice		
One month or less (N=156)	31	
More than one month (N=140)	21	
Perception of notice adequacy		
Adequate (N=133)	23	
Insufficient (N=55)	38	
Family readiness		
Ready or very ready (N=214)		32
Somewhat ready (N=70)		56
Not at all ready (N=55)		51

SOURCE: 2006 RAND Guard and Reserve Family Interviews.

NOTES: Ns are provided for either service member or spouse, as denoted in the table. When data from both groups are shown, Ns are specified as service member or spouse. All percentages shown are statistically different from one another at $p<0.10$. Shaded cells indicate subsets of the population that are not significantly different from other subsets. For reserve component comparisons among service members, only the Army National Guard is significantly different from the Army Reserve. For reserve component comparisons among spouses, the Air Force Reserve is significantly different from the Army Reserve and from the Marine Forces Reserve. The other component comparisons are not significantly different.

Among spouses, younger spouses, those who had been married for less time, and those who were not parents were more likely to mention emotional or mental problems. Their past experience with the military also made a difference: Spouses who had previously served in the military and spouses married to service members that had previously served in the active component were less likely to mention emotional or mental problems. Presumably they were either generally more accustomed to these stressors or they did not experience them to the same degree. Additionally, the extent to which the family was ready also helped to explain differences in responses, in that spouses from less-ready families were more likely to experience emotional problems.

There were also some differences among respondents by their reserve component. Specifically, Air Force Reserve spouses were less likely to note emotional problems than were the spouses of the Army Reserve or Marine Forces Reserve. Among service members, Army guardsmen were less likely than Army reservists to mention such problems.

Household Responsibility Problems

While about 20 percent of service members acknowledged difficulty in the household while they were gone, twice that proportion of spouses mentioned problems accommodating the demands of family life, to include comments about child care, household chores, and chauffeuring children, such as the following:

> More responsibility fell on me to get things done. Before I never mowed a yard and now I had to while he was gone. So that was an experience for me; I never had to do that before. You know, taking out the trash, had to do all that. (100: Army National Guard, E-6's wife with three children)

> The fact that my wife, with me being gone, had to deal with making sure the bills are being paid, taking care of the yard, chores around the house, all the things I would have done or helped doing, she had to take on all by herself. (178: Marine Forces Reserve, O-3 without children)

> The household issues as far as the yard work and trying to keep the house clean and you don't have an extra set of hands to help you out. When your child gets sick and they have to be pulled out of day care, there's only one parent that can stay home with them. (449: Army National Guard, E-5's wife with one child)

As these comments illustrate, the issues ranged from having to take the trash out and the difficulties spouses had learning to pay the monthly bills, to more serious home issues, such as cleanup after severe storm damage and, for a few spouses, caring for the family farm:

We had some household emergencies while I was deployed . . . because a tree hit the house. (177: Air Force Reserve, E-6)

With a 19-month-old, and she was only ten months when my husband left. I guess that's the biggest challenge, is trying to parent her on my own and have a full-time job, and we live on a farm, so I'm trying to take care of that. We rent the land out, so that is taken care of, but I've got the responsibility of the fertilizer and chemicals, and just paying the bills, and everything falls on me. (259: Army National Guard, E-6's wife with one child)

As shown in Table 4.2, there were few patterns that distinguished the service members who were more likely to mention household responsibilities. Service members who lived far from the nearest military installation mentioned this problem more than those who lived closer. Additionally, Marine reservists were less likely than Army guardsmen to mention these issues. There were also differences by component among the spouses interviewed: Air Force Reserve and Marine Forces Reserve spouses were less likely than were Army Guard and Army Reserve spouses to mention problems satisfying household responsibilities. There were other patterns evident in the spouse portion of the sample. Female spouses, spouses who were parents, and spouses who had been married longer were more likely to mention these issues, likely reflecting the greater complexity of well-established households. Indeed, almost half of spouses who were parents mentioned such issues. Those spouses who experienced longer deployments and those who lived farther from the nearest military installation were more likely to mention these issues. Forty-six percent of spouses whose service member was deployed one year or longer mentioned these issues, compared with 35 percent of spouses experiencing shorter deployments. We also found that spouses from families in more comfortable financial situations were less likely to cite these problems. In fact, approximately half of spouses that described their current financial situation as anything other than comfortable experienced problems in the household, suggesting that financial resources may ameliorate some of these challenges.

Table 4.2
Characteristics Associated with Citing Household Responsibility Issues

	Service Members (%)	Spouses (%)
Overall percentage citing household responsibility issues (N=296 service members; N=357 spouses)	20	40
Gender		
Male (N=12)		17
Female (N=345)		41
Marriage length		
2 years or less (N=71)		28
3 years or more (N=286)		43
Parental status		
Has children (N=269)		46
No children (N=88)		24
Service member reserve component		
Army National Guard (N=104 service members; N=102 spouses)	25	48
Army Reserve (N=74 service members; N=89 spouses)	18	48
Air Force Reserve (N=60 service members; N=83 spouses)	23	33
Marine Forces Reserve (N=58 service members; N=83 spouses)	12	30
Financial situation		
Comfortable (N=255)		36
Occasional difficulty (N=74)		54
Uncomfortable (N=26)		50
Distance from nearest military installation		
Less than 100 miles (N=247 service members; N=292 spouses)	18	38
100 miles or more (N=49 service members; N=31 spouses)	31	65

Table 4.2—Continued

	Service Members (%)	Spouses (%)
Deployment length		
Less than one year (N=180)		35
One year or more (N=169)		46

SOURCE: 2006 RAND Guard and Reserve Family Interviews.

NOTES: Ns are provided for either service member or spouse, as denoted in the table. When data from both groups are shown, Ns are specified as service member or spouse. All percentages shown are statistically different from one another at p<0.10. Shaded cells indicate subsets of the population that are not significantly different from other subsets. For reserve component comparisons among service members, the Army National Guard is significantly different from the Marine Forces Reserve. For reserve component comparisons among spouses, the Air Force Reserve is significantly different from both the Army National Guard and the Army Reserve. The Marine Forces Reserve is significantly different from both the Army National Guard and the Army Reserve. The other component comparison is not significantly different.

Children's Issues

Approximately 26 percent of the spouses and about half that proportion of the service members interviewed mentioned concerns about the effect of the deployment on their children. These effects included a range of emotional or mental problems, as well as other sacrifices or difficulties experienced by children. In general, we were inclusive when we assessed these data. In other words, while we did not include spouses who simply missed their service members among spouses with emotional issues, we did, in this instance, include cases of parents talking about how much their children missed their uniformed parent. Our logic was that we, and perhaps even the parents, are unable to assess the impact that living without a deployed parent for an extended period of time may have on a child. Other sacrifices or issues were more apparent, such as children who were unable to participate in their usual extracurricular programs, very young children who seemed not to recognize their parent upon return, and children who developed behavioral or academic problems during the deployment. Comments included the following:

There are some problems with the kids, that they are facing, attachment issues. . . . We have a two-year-old who wouldn't sleep in his own bed for the first seven months that he was gone. He wouldn't let me out of his sight. He was scared I was going to leave. Our ten year old stopped joining sports, he just wanted to stay home. Stuff like that. (5: Army National Guard, E-5's wife with three children)

With my child, when I got back I just had to deal with the whole him getting used to me and remembering who I was. Then I moved from New York to California, and I had to basically drag him out from the family that he had been for a whole year and bring him with me. I might have seemed like a stranger to him. He was over one and a half. (536: Marine Forces Reserve, female E-4 with two children)

The kids, every day asking for daddy, that's hard. When they constantly—They don't understand. They're four and six, they don't get it. My youngest asks all the time if daddy is going to die. (597: Army Reserve, E-4's wife with two children)

We've had some problems with the kids, mostly my older one getting in trouble where he never has before. (106: Army Reserve, E-6's wife with three children)

My teenage children claim that their social life was destroyed. Because Mom was too tired to drive them somewhere. My daughter turned 16 when he was gone and she is 17 now and still doesn't have her driver's license because we didn't have the time to take her out as much as she needed to before her test. (87: Army National Guard, E-4's wife with two children)

Not surprisingly, as Table 4.3 indicates, children's issues were mentioned predominantly by spouses and service members that were parents. The likelihood that service members interviewed would mention children's issues was higher for female service members and those who had experienced multiple deployments, as well as for those that had received a month or less of deployment notice. Among spouses, older spouses and spouses who had been married longer were also more likely to be concerned about the effect on children. Service member component and pay grade also served as a basis for differences in responses, as 37 percent of junior enlisted spouses mentioned children's issues, compared with only approximately 23 percent of the other spouses interviewed, and Army Guard spouses were more likely than were Marine Forces Reserve spouses to mention children's issues. Spouses who felt their family had received insufficient notice of the deployment or who experienced a deployment of one year or longer were also more likely to discuss children's issues. Additionally, almost half of spouses who reported uncomfortable financial circumstances mentioned this issue, compared with about one-fifth of those enjoying comfortable financial circumstances.

Financial and Legal Problems

Although the military family experts we interviewed tended to emphasize the financial problems that accompany deployment, only 15 percent overall of service members and spouses mentioned a financial or legal problem. Further, as will be discussed in the next chapter, financial *gain* was a frequently mentioned as a positive aspect of deployment.[2] This relatively small proportion of interviewees that felt financial issues were problematic during deployment is consistent with the other findings suggesting that many deployed personnel enjoy financial gain (Loughran, Klerman, and Martin, 2006; Hosek, Kavanagh, and Miller, 2006).

Most of the comments from those who did cite financial problems referred to issues such as accommodating the gap in pay until the family received the first military paycheck. Some personnel did refer to lost income, and a handful of our sample reported losses they regarded as significant, either during the deployment or post-deployment. Among the service members experiencing post-deployment financial problems

---

[2] We considered financial problems to include issues about having sufficient funds, financial planning, difficulty with the military pay system, etc. Comments about paying the bills that referred to the mechanics of bill paying, such as writing a check and mailing the envelope, were included in household responsibilities.

Table 4.3
Characteristics Associated with Citing Children's Issues

	Service Members (%)	Spouses (%)
Overall percentage citing children's issues (N=296 service members; N=357 spouses)	12	26
Age		
25 or less (N=55)		16
26 or more (N=302)		28
Gender		
Male (N=270 service members; N=12 spouses)	10	
Female (N=26 service members; N=345 spouses)	23	
Marriage length		
2 years or less (N=71)		15
3 years or more (N=286)		29
Parental status		
Has children (N=232 service members; N=269 spouses)	14	34
No children (N=64 service members; N=88 spouses)	3	3
Service member pay grade		
E-1 to E-4 (N=90)		37
E-5 to E-6 (N=174)		24
O-1 to O-3 (N=93)		22
Service member reserve component		
Army National Guard (N=104 service members; N=102 spouses)		31
Army Reserve (N=74 service members; N=89 spouses)		30
Air Force Reserve (N=60 service members; N=83 spouses)		22
Marine Forces Reserve (N=58 service members; N=83 spouses)		20

Table 4.3—Continued

	Service Members (%)	Spouses (%)
Financial situation		
Comfortable (N=255)		22
Occasional difficulty (N=74)		35
Uncomfortable (N=26)		46
Amount of notice		
One month or less (N=156 service members)	15	
More than one month (N=140 service members)	8	
Perception of notice adequacy		
Adequate (N=133 service members; N=161 spouses)		25
Insufficient (N=55 service members; N=62 spouses)		42
Deployment length		
Less than one year (N=180)		23
One year or more (N=169)		31
Repeat OCONUS deployments		
Yes (N=48)	21	
No (N=248)	10	

SOURCE: 2006 RAND Guard and Reserve Family Interviews.

NOTES: Ns are provided for either service member or spouse, as denoted in the table. All percentages shown are statistically different from one another at $p<0.10$. Shaded cells indicate subsets of the population that are not significantly different from other subsets. For reserve component comparisons among spouses, the Army National Guard is significantly different from the Marine Forces Reserve. The other component comparisons are not significantly different. For pay grade comparisons among the spouses, the E-1 to E-4 category differs significantly from both the E-5 to E-6 and the O-1 to O-3 categories, but the other pay grade comparison is not significantly different.

were those with PTSD symptoms so severe they were unable to work when they returned home.[3] Comments included the following:

> Financial was a big thing. Getting the Iraqi phone cards was a financial burden that we had on the family. Being able to communicate was very important to us. No matter what the cost was, we wanted to talk as much as we could. (115: Army National Guard, E-6's wife)

> The biggest issue we've had is from the last civilian paycheck to the first military paycheck. Just knowing, before you used to get the paycheck every 2 weeks, is it going to be 3 weeks before I get one, is going to be 4 weeks before I get one or is it going to be next week before I get one. It's all set on a calendar so we know this, we know the answer. It's just the financially getting the bills on time and getting your financing rescheduled for that difference in pay dates. (361: Air Force Reserve, female E-5)

> Financial ruin. He came home in June, was diagnosed with PTSD. There is also something physically wrong with him and the doctors don't know what it is. He is depressed and lost; he is on medication and is not back to work. It's taken a toll on him and, in turn, it's taken a toll on our family. The VA [Veterans Affairs] is no help, the public sector is no help as far as the employment office and companies that are hiring. He went over there [to Iraq] and almost died and nobody will hire him. Someone will shake your hand and thank you [for fighting in Iraq] but that's as far as it goes. (902: Army National Guard, E-6's wife)

Those comments referring to legal issues included mentions of the difficulties closing on a new home or pursuing adoption while the service member was deployed, as well as some child custody issues that were complicated by the deployment.

---

[3]   We had only very small numbers of returned service members who mentioned PTSD, but we include it here because related research found that approximately 21 percent of National Guard and Reserve component service members screen positive for PTSD, major depression, or other mental health problems (Hoge, Auchterlonie, and Milliken, 2006.)

Significant differences in the characteristics of interviewees who discussed financial and legal problems are shown in Table 4.4. Not surprisingly, financial and legal issues were most frequently discussed by interviewees that described their current financial situation as anything other than comfortable. While only about 10 percent of spouses and service members that described their financial situation as comfortable mentioned financial or legal difficulties, about 40 percent of interviewees in an uncomfortable financial situation mentioned such problems. Other patterns that were significant among both service members and spouses include their perception of whether they had adequate notice of deployment and whether their family had been ready for the deployment.

Among the service members, there were some differences by reserve component; Army guardsmen were less likely to mention financial or legal problems than were Army reservists. Service members who lived farther from their drill unit and service members who lived farther from the nearest military installation were also more likely to describe these problems.

Younger spouses were less likely to mention these problems, and spouses married to junior officers were less likely to mention these issues than were spouses married to enlisted personnel. Also, spouses who had received one month or less of notice (in addition to those who felt their notice was inadequate) were more likely to mention financial or legal issues.

Employment and Education Problems

This research analyzed the extent to which interviewees mentioned employment or education problems when asked broadly about the issues resulting from deployment. In addition to this question, on which most of the findings in this chapter are based, we specifically asked reservists and guardsmen about the effect of their Reserve or National Guard service on their education or employment, and we asked spouses about the extent to which their employers or coworkers were supportive during their service member's deployment.

Table 4.4
Characteristics Associated with Citing Financial or Legal Problems

	Service Members (%)	Spouses (%)
Age		
25 or less (N=55)		7
26 or more (N=302)		18
Service member pay grade		
E-1 to E-4 (N=90)		17
E-5 to E-6 (N=174)		21
O-1 to O-3 (N=93)		8
Service member reserve component		
Army National Guard (N=104)	8	
Army Reserve (N=74)	23	
Air Force Reserve (N=60)	13	
Marine Forces Reserve (N=58)	16	
Financial situation		
Comfortable (N=195 service members; N=255 spouses)	7	11
Occasional difficulty (N=66 service members; N=74 spouses)	21	24
Uncomfortable (N=35 service members; N=26 spouses)	40	42
Distance from drill unit		
Less than 100 miles (N=218)	11	
100 or more miles (N=78)	23	
Distance from nearest military installation		
Less than 100 miles (N=247)	12	
100 or more miles (N=49)	27	
Amount of notice		
One month or less (N=157)		20
More than one month (N=189)		13
Perception of notice adequacy		
Adequate (N=133 service members; N=161 spouses)	14	12
Insufficient (N=55 service members; N=62 spouses)	25	21

Table 4.4—Continued

	Service Members (%)	Spouses (%)
Family readiness		
Ready or very ready (N=192 service members; N=214 spouses)	9	12
Somewhat ready (N=43 service members; N=70 spouses)	26	21
Not at all ready (N=51 service members; N=55 spouses)	22	24

SOURCE: 2006 RAND Guard and Reserve Family Interviews.

NOTES: Ns are provided for either service member or spouse, as denoted in the table. When data from both groups are shown, Ns are specified as service member or spouse. All percentages shown are statistically different from one another at p<0.10. Shaded cells indicate subsets of the population that are not significantly different from other subsets. For reserve component comparisons among service members, the Army National Guard is significantly different from the Army Reserve. The other component comparisons are not significantly different. For pay grade comparisons among the spouses, the O-1 to O-3 category differs significantly from the E-1 to E-4 and the E-5 to E-6 categories. The other pay grade comparison is not significantly different.

When asked directly about the effect of their reserve component service on work or education, 47 percent of service members indicated an effect of some type on education or employment. Of those who mentioned an effect, roughly three-fourths cited a negative effect, and approximately one-fourth referred to a positive effect. When discussing more broadly all the kinds of problems encountered during or resulting from the most recent deployment, 12 percent of spouses and 21 percent of service members mentioned employment problems. This reporting difference between spouses and service members is partly explainable by the fact that some of the employment issues emerged for families after the deployment and, while all the interviewed reservists and guardsmen had returned from deployment, some of the spouses still had deployed service members. Across both questions asking about employment problems, responses included references to lost jobs, missed career opportunities, and lapsed or outdated expertise. Some service members and spouses reported that employers did not retain the service member's job (in spite of this being illegal) or placed them in a job that the service member perceived to be a lesser opportunity. Still other reservists and guardsmen mentioned lost opportunities for professional development. Comments included

> His civilian employer was very uncooperative. My husband's employer is not supportive of any military members. When he had returned from being overseas, they tried to fire him. (231: Air Force Reserve, E-5's wife)

> My employment when I came back, they didn't really give me my job back that I had before. So it's been difficult. . . . I'm getting the same pay, but my duties are a lot less than what they used to be. (65: Marine Forces Reserve, female E-5)

> He applied for the DEA [Drug Enforcement Administration] and it is a one-year process. He got almost halfway into it and he got activated so he is going to have to restart the whole process when he gets back. (700: Marine Forces Reserve, O-3's wife)

> From the standpoint of my ability to go back to my civilian job, it has really come to a negative effect because I had just invested over $4,000 to get a commercial license to drive a truck over the railroad, and then two months later I got activated for two years. And when I got activated again last year, due to an injury, I now cannot drive a tractor trailer. So I cannot go back to work in my original employment. Once I am released medically, I'm going to have to find a different line of work. So I would say it is a negative effect because I had invested money to gain sufficient civilian employment and licensing, but now I cannot use that. (257: Air Force Reserve, E-6)

> It put me back in my technology. I'm a software engineer. I was gone for approximately two years, and the two years I was gone, my company moved from a lower platform to a higher platform, which would be Microsoft.net and during that transition, in two years I fell behind in my skill set as a software engineer. (574: Army Reserve, O-2)

The comments from interviewees referred not only to problems with the service member's employment, as included above, but also problems with the spouse's employment, such as employers that would not or could not accommodate the flexibility needed by the spouse during the deployment:

> I would say probably the loss of my job. It just became so difficult to try and find, to try and work out a schedule with my employer that would accommodate him [her service member] not being around to help with the children. (134: Air Force Reserve, E-6's wife with two children)

> For prospective employers I'm being told that now that my husband is being deployed, I'm a single parent and they don't want to hire me because of that, because of the stress that they say I'm under. (142: Army National Guard, E-4's wife with one child)

> Her job was affected because of my mobilization, and her not having the latitude and the flexibility of her schedule if I was around. With me being gone she didn't have the latitude to travel as much for work and it kind of hindered her career as well. (418: Army Reserve, O-3 with two children)

This is particularly notable because roughly half of both employed spouses and service members who reported that their spouse was employed (53 percent and 49 percent, respectively) indicated that they (their spouse) made either a moderate or a major contribution toward the family's monthly household income. In all, however, spouses generally reported that their own employers and coworkers were supportive during the deployment, with the majority of spouses saying that both their employers (80 percent) and coworkers (91 percent) were supportive, and with only 8 percent of spouses claiming that their employer was unsupportive, and even fewer (2 percent) claiming that coworkers were unsupportive.

Table 4.5 depicts the characteristics of the interviewees that discussed employment problems resulting from the deployment. Among service members, those most likely to mention employment problems

Table 4.5
Characteristics Associated with Citing Employment Problems

	Service Members (%)	Spouses (%)
Overall percentage citing employment problems (N=296 service members; N=357 spouses)	21	12
Age		
25 or less (N=27)	5	
26 or more (N=258)	22	
Marriage length		
2 years or less (N=71)	8	
3 years or more (N=286)	24	
Parental status		
Has children (N=232)	23	
No children (N=64)	13	
College degree		
Yes (N=125 service members; N=195 spouses)	26	17
No (N=172 service members; N=162 spouses)	17	4
Service member pay grade		
E-1 to E-4 (N=69)	16	
E-5 to E-6 (N=146)	19	
O-1 to O-3 (N=81)	27	
Service member reserve component		
Army National Guard (N=104)	18	
Army Reserve (N=74)	35	
Air Force Reserve (N=60)	10	
Marine Forces Reserve (N=58)	17	
Distance from nearest military installation		
Less than 100 miles (N=247)	19	
More than 100 miles (N=49)	31	

Table 4.5—Continued

	Service Members (%)	Spouses (%)
Amount of notice		
One month or less (N=156)	26	
More than one month (N=140)	14	

SOURCE: 2006 RAND Guard and Reserve Family Interviews.

NOTES: Ns are provided for either service member or spouse, as denoted in the table. When data from both groups are shown, Ns are specified as service member or spouse. All percentages shown are statistically different from one another at p<0.10. Shaded cells indicate subsets of the population that are not significantly different from other subsets. For pay grade comparisons among service members, the E-1 to E-4 category differs significantly from the O-1 to O-3 category. The other pay grade comparisons are not significantly different. For reserve component comparisons among service members, the Army National Guard differs significantly from the Army Reserve. The other component comparisons are not significantly different.

included older service members, parents, those past the newlywed stage, and those that were college-educated. There were also some pay grade and component differences among the service members, in that junior officers were more likely to mention employment problems than were junior enlisted personnel, and Army reservists were considerably more likely to do so as well. Service members who resided farther from the nearest military installation were also more likely to cite employment problems, as were those who received one month or less of deployment notice.

There were few patterns among the spouses regarding those who noted employment problems, although college-educated spouses were more likely to note such problems.

About 15 percent of service members responded to the aforementioned direct question about the effect of their reserve component service on work or education by noting a negative effect on their education. A similar proportion included education problems in their response to the broader question about issues and problems faced as a result of their deployment.

As Table 4.6 shows, when asked broadly about the problems incurred as a result of their deployment, younger service members, female service members, and junior enlisted personnel were more likely

Table 4.6
Characteristics Associated with Citing Education Problems

	Service Members (%)	Spouses (%)
Overall percentage citing education problems (N=296 service members; N=357 spouses)	14	4
Age		
25 or less (N=27)	24	
26 or more (N=258)	12	
Marriage length		
2 years or less (N=71)		9
3 years or more (N=286)		3
Gender		
Male (N=270)	12	
Female (N=26)	35	
Service member pay grade		
E-1 to E-4 (N=69)	22	
E-5 to E-6 (N=146)	13	
O-1 to O-3 (N=81)	9	
Service member reserve component		
Army National Guard (N=104 service members; N=102 spouses)	11	1
Army Reserve (N=74 service members; N=89 spouses)	15	3
Air Force Reserve (N=60 service members; N=83 spouses)	12	4
Marine Forces Reserve (N=58 service members; N=83 spouses)	21	8
Amount of notice		
One month or less (N=156)	17	
More than one month (N=140)	10	
Perception of notice adequacy		

Table 4.6—Continued

	Service Members (%)	Spouses (%)
Adequate (N=133)	10	
Insufficient (N=55)	24	
Repeat OCONUS deployments		
Yes (N=132)		15
No (N=225)		9

SOURCE: 2006 RAND Guard and Reserve Family Interviews.

NOTES: Ns are provided for either service member or spouse, as denoted in the table. When data from both groups are shown, Ns are specified as service member or spouse. All percentages shown are statistically different from one another at p<0.10. Shaded cells indicate subsets of the population that are not significantly different from other subsets. For pay grade comparisons among service members, the E-1 to E-4 category differs significantly from the O-1 to O-3 category. The other pay grade comparisons are not significantly different. For reserve component comparisons among service members and among service members, the Army National Guard differs significantly from the Marine Forces Reserve. The other component comparisons are not significantly different.

than their counterparts to refer to education problems. There were also some patterns by component, in that Marine reservists were more likely than were Army guardsmen to mention education problems. Additionally, service members with less notice and those who believed their notice to have been inadequate were also more likely to mention education problems.

Among spouses, those who were newly married were more likely to mention education problems. The component pattern evident among service members was also evident for spouses, albeit at much lower levels: Those married to Marine reservists were slightly more likely than those married to Army guardsmen to note education problems. Additionally, spouses who had experienced multiple deployments were also more likely to mention education problems.

Marital Problems

As shown earlier, 12 percent of interviewees reported marital problems as a result of the deployment. A small number of these cases involved a recent or an impending divorce. While some comments simply mentioned "marital strife" or "marital problems," others spoke broadly of the difficulty maintaining a marriage from a distance:

> Well, we faced issues relating to the difficulty of being away from each other for that long. Being married with no children was an advantage in that our children didn't have to go through it, but it was a disadvantage in that there was no tie other than the marriage that we agreed to. And, of course, rumors that had emerged from acquaintances around her made things difficult. It put stresses into her life related to our marriage and my fidelity and all kinds of silly things, and so we had to deal with that from 10,000 miles away over a phone with a four or five second delay. It was challenging. (572: Army Reserve, O-3)

> Marriage wise, I would say I can anticipate us probably being in counseling when this is all over. There's always going to be problems and issues, and you can't blame it all on the deployment. But, it is a huge contributing factor. (452: Army National Guard, O-3's wife)

Others spoke more specifically of the difficulty readjusting to one another after the deployment, as in these instances:

> I did everything by myself while he was gone and I expect him to pick right back up and do everything he did before. But because he's only taken care of himself for the last 18 months he doesn't necessarily—we're not on the same sheet as far as that's concerned. So I think probably trying to make it a 50/50 partnership again is difficult. (226: Army National Guard, E-5's wife)

> Really, one of the worst things for that is that my wife controls all the money and all the bills and is the rule maker for the children. Then when I come back, I have to ask her to step down and let me

assume my role again. I think that might be the worst worse part
of the activation. (97: Army National Guard, E-6)

I'd say the main thing was marital problems, and that's just from
going to a co-dependency state to an independent state and then
back again to a co-dependent state. (629, Air Force Reserve,
female E-4)

Table 4.7 indicates the differences in characteristics among interviewees who mentioned marital problems. Female service members
were more likely to note marital problems upon their return, as were
service members with less comfortable family finances. Among spouses,
officers' spouses and spouses who had experienced longer deployments
were more likely to mention marital problems. Likewise, both spouses
and service members who felt that their family was anything other
than ready or very ready for their deployment were more likely to mention marital problems.

Health Care Problems

The military family experts interviewed at the beginning of our study
were inclined to mention health care problems as a predominant issue
for guard and reserve families, but health care problems were reported
by only roughly one-tenth of both spouses and service members interviewed. As would be expected based on the expert interviews, these
service members and spouses provided comments referring to the difficulty in identifying doctors that would accept TRICARE, medical
claim issues, and disappointment over what would be covered, either
through TRICARE or, in the case of wounded service members,
through the VA:

TRICARE, TRICARE, TRICARE. If you live beyond the con-
fines of an active duty post the medical facilities are not familiar
with TRICARE and TRICARE is not familiar with the community, they don't know what doctors are there, what doctors
are available, what doctors the families are using. I still have over

Table 4.7
Characteristics Associated with Citing Marital Problems

	Service Members (%)	Spouses (%)
Gender		
Male (N=270)	11	
Female (N=26)	23	
Service member pay grade		
E-1 to E-4 (N=90)		7
E-5 to E-6 (N=174)		10
O-1 to O-3 (N=93)		16
Financial situation		
Comfortable (N=195)	10	
Occasional difficulty (N=66)	14	
Uncomfortable (N=35)	23	
Deployment length		
Less than one year (N=180)		7
One year or more (N=169)		15
Family readiness		
Ready or very ready (N=192 service members; N=214 spouses)	7	7
Somewhat ready (N=43 service members; N=70 spouses)	19	19
Not at all ready (N=51 service members; N=55 spouses)	27	16

SOURCE: 2006 RAND Guard and Reserve Family Interviews.

NOTES: Ns are provided for either service member or spouse, as denoted in the table. When data from both groups are shown, Ns are specified as service member or spouse. All percentages shown are statistically different from one another at $p<0.10$. Shaded cells indicate subsets of the population that are not significantly different from other subsets. For pay grade comparisons among spouses, the E-1 to E-4 category differs significantly from the O-1 to O-3 category. The other pay grade comparisons are not significantly different.

$700 in medical issues that happened while I was deployed that I can't seem to get rectified. (203: Army National Guard, O-3)

The VA keeps denying me what I should get [for] my teeth, because I had some teeth knocked out and they keep denying me my claim on that. I hurt my back in Iraq, they denied my claim on that. You go over and fight a war for these people and they deny you your health care for it. (170: Army Reserve, E-6)

I know a big one was health insurance. Just the way the military has set it up, it is geared [more] for active duty than it is for reservist. . . . It is geared for big cities. We live in a rural area and I know that the medical part of it wanted us to drive more than two hours away for a regular doctor. She had to fight that. She finally got that situated that she went back to our regular doctor. I would say that was probably the biggest one there. (469: Army Reserve, E-5 with two children)

My family moved about 100 miles during the first deployment, and finding new doctors under TRICARE in the area was a little bit of a problem. . . . We live really close to the state line, and the health care people kept trying to send us like 80 miles to a doctor, because they wouldn't look in a different state than the state we lived in, even though that was closer for us. (570: Army Reserve, E-6 with four children)

However, the small number of spouses and service members who discussed health care–related challenges may suggest either that health care issues have improved since we initially interviewed guard and reserve family experts (five to six months earlier), or that problems are not widespread. We did not find many significant demographic differences among the service members that reported these problems, although there were some differences in the spouse portion of our interview sample, as shown in Table 4.8. Specifically, patterns based on reserve component and family readiness were present; spouses of Army reservists were more likely to mention health care problems than were those of Air Force reservists. Also, spouses who characterized their family as ready or very ready were less inclined to discuss problems of this nature.

Table 4.8
Characteristics Associated with Citing Health Care Problems

	Service Members (%)	Spouses (%)
Service member reserve component		
Army National Guard (N=102)		12
Army Reserve (N=89)		15
Air Force Reserve (N=83)		6
Marine Forces Reserve (N=83)		13
Family readiness		
Ready or very ready (N=214)		7
Somewhat ready (N=70)		24
Not at all ready (N=55)		13

SOURCE: 2006 RAND Guard and Reserve Family Interviews.

NOTES: Ns are provided for either service member or spouse, as denoted in the table. All percentages shown are statistically different from one another at p<0.10. Shaded cells indicate subsets of the population that are not significantly different from other subsets. For reserve component comparisons among spouses, the Army Reserve differs significantly from the Air Force Reserve. The other differences are not statistically significant.

No Problems

Even after prompting interview participants by reading typical problems, 29 percent of service members maintained that their family had not experienced problems as a result of their deployment. Only 14 percent of spouses made this same assertion.[4]

Table 4.9 indicates several patterns that were evident for both service members and spouses who stated that their family had not experienced any problems. Among both groups, interviewees in more comfortable financial situations, those who felt they had received adequate notice, and those who felt that their family had been either ready or very ready for the deployment were most likely to say they had not experienced problems. Among service members, Army reservists were

---

[4] This includes only the interview participants that claimed they did not have problems. There was a additional small number of participants who declined to or otherwise did not answer that question.

Table 4.9
Characteristics Associated with Citing No Problems

	Service Members (%)	Spouses (%)
Overall percentage citing no problems (N=296 service members; N=357 spouses)	29	14
Gender		
Male (N=12)		42
Female (N=345)		13
Parental status		
Has children (N=269)		12
No children (N=88)		20
Service member reserve component		
Army National Guard (N=104 service members; N=102 spouses)	32	8
Army Reserve (N=74 service members; N=89 spouses)	18	9
Air Force Reserve (N=60 service members; N=83 spouses)	32	24
Marine Forces Reserve (N=58 service members; N=83 spouses)	38	17
Spouse prior military (spouses only)		
Yes (N=34)		26
No (N=323)		13
Financial situation		
Comfortable (N=195 service members; N=255 spouses)	37	17
Occasional difficulty (N=66 service members; N=74 spouses)	17	7
Uncomfortable (N=35 service members; N=26 spouses)	11	4
Distance from drill unit		
Less than 25 miles (N=94)		20
25 miles or more (N=228)		11

Table 4.9—Continued

	Service Members (%)	Spouses (%)
Perception of notice adequacy		
Adequate (N=133 service members; N=161 spouses)	35	17
Insufficient (N=55 service members; N=62 spouses)	9	3
Deployment length		
Less than one year (N=180)		18
One year or more (N=169)		10
Family readiness		
Ready or very ready (N=192 service members; N=214 spouses)	39	19
Somewhat ready (N=43 service members; N=70 spouses)	16	1
Not at all ready (N=51 service members; N=55 spouses)	10	7

SOURCE: 2006 RAND Guard and Reserve Family Interviews.

NOTES: Ns are provided for either service member or spouse, as denoted in the table. When data from both groups are shown, Ns are specified as service member or spouse. All percentages shown are statistically different from one another at p<0.10. Shaded cells indicate subsets of the population that are not significantly different from other subsets. For reserve component comparisons among service members, the Army Reserve differs significantly from the Air Force Reserve and from the Marine Forces Reserve. For reserve component comparisons among spouses, the Army Guard differs significantly from the Air Force Reserve and from the Marine Forces Reserve; the Army Reserve differs significantly from the Air Force Reserve. The other differences are not statistically significant.

the least likely to claim that they had no problems. Husbands of service members, spouses without children, and spouses married to Air Force reservists were most likely to assert "no problems." In addition, spouses who had prior military experience were more likely to assert that they had no problems, as were spouses who lived closer to the drill unit and those who experienced deployments of one year or less.

Discussion

When we discussed with the military family experts the challenges that guard and reserve families face, the majority of the experts mentioned health care issues, emotional problems, household responsibility issues, and financial problems. Our interviews with spouses and service members corroborate some, but not all, of the experts' perceptions of the problems facing guard and reserve families. The majority of both service members and spouses included in our study did mention problems. Household responsibility problems, emotional or mental problems, and children's issues were the top three problems most frequently mentioned by spouses, whereas the top three most-frequently mentioned problems by service members were emotional or mental problems, employment problems, and household responsibility problems. None of these problems were cited by a majority of the interview participants, be they service members or spouses. Table 4.10 summarizes the relationships between individual and situational characteristics and the problems discussed in this chapter. The table also denotes whether the pattern was evident in the service member or spouse portion of the interview sample. For example, age was related to a tendency among spouses to cite emotional or mental problems, children's issues, and financial and legal issues, as well as a tendency among service members to cite employment and education issues.

Both the experts we interviewed and prior research (e.g., Caliber Associates, 2003) identified junior enlisted and younger families as more vulnerable, or more likely to experience problems. We found in our research that different families have different kinds of problems; certain characteristics are more likely to be associated with some of the problems and less likely to be associated with others. For example, as noted above, the age of spouses was a factor in whether they tended to report some problems. But the relationship differed: Younger spouses (age 25 and under) and those newer to marriage were more likely to report emotional or mental problems, whereas older spouses were more likely to mention household responsibility problems, children's issues, and financial and legal problems. Similarly, older personnel were more likely to be concerned about employment problems, whereas younger

Table 4.10
Summary of Factors Related to Reported Problems

	Emotional or Mental	Household Respon- sibilities	Children's Issues	Financial or Legal	Employment	Education	Marital	Health Care	No Problems
Individual and situational characteristics									
Age	SP		SP	SP	SM	SM			
Gender		SP	SM			SM	SM		SP
Marriage length	SP	SP	SP		SM				
Parental status	SP	SP	SM, SP		SM				SP
College degree					SM, SP				
Service member pay grade			SP	SP	SM	SM	SP		
Service member reserve component	SM, SP	SM, SP	SP	SM	SM	SM, SP		SP	SM, SP
Service member prior active duty	SP								
Spouse prior military (spouses only)	SP								SP

Table 4.10—Continued

	Emotional or Mental	Household Responsibilities	Children's Issues	Financial or Legal	Employment	Education	Marital	Health Care	No Problems
Financial situation	SM	SP	SP	SM, SP			SM		SM, SP
Distance from drill unit				SM					SP
Distance from nearest military installation		SM, SP		SM	SM				
Amount of notice	SM		SM	SP	SM	SM			
Perception of notice adequacy	SM		SP	SM, SP		SM			SM, SP
Deployment length		SP	SP				SP		SP
Repeat OCONUS deployments			SM			SP			
Family readiness	SP			SM, SP			SM, SP	SP	SM, SP

SOURCE: 2006 RAND Guard and Reserve Family Interviews.

NOTES: All relationships listed are statistically significant at p<0.10. SM = Finding present in the service member portion of the sample (N=296); SP = Finding present in the spouse portion of the sample (N=357).

personnel were more focused on education problems. Where gender was a factor for the problems, women (both spouses and service members) were more likely to report the problem (or less likely to say "no problem.") However, since these findings stem from self-reported data, it is unclear whether women experience more problems or whether women are more inclined to mention them. Current family finances often factored into the problems mentioned, as did the extent to which the family had been ready for the deployment. Prior military experience was a factor in only a couple instances: Spouses who had formerly served in the military were less likely to mention emotional or mental problems and were more likely to report no problems than were other spouses, and the spouses of service members who had previously served on active duty were also less likely to note emotional or mental problems. Extreme distance from the nearest military installation accounted for a tendency to identify household responsibility problems for both spouses and service members, as well as financial and legal issues and employment issues for service members. Proximity to the drill unit was related to two patterns: Service members who lived very far from the drill unit were more likely to cite financial problems, and spouses who lived close to the drill unit were more likely to assert that they had no problems.

In addition, interview participants who received less notice or who believed they had received inadequate notice of deployment were more likely to mention most of the problems discussed. Related to that, service members and spouses who described their family as unprepared for the deployment were also more likely to report some of the problems, although we do not know whether the presence of such problems in their family affected their retrospective determination of whether they had been ready.

Finally, we note that even after additional prompting, 29 percent of service members maintained that their family had not experienced any problems. Fourteen percent of spouses made this assertion. While this difference might be attributed to the units from which participants were selected, we cannot confirm that. Thus, we must instead question whether service members are either less cognizant of challenges faced by their families during deployment, are less inclined to identify those issues, or whether, given that the service members interviewed were predominantly male, this is a reflection of the gender issue discussed above.

CHAPTER FIVE

What Positives Do Guard and Reserve Families Report?

This research effort also considered the positives experienced by reserve component families as a result of activation or deployment. When we asked the experts on guard and reserve family issues to discuss the positives that they felt these families incurred, patriotism and personal gratification were the most frequently mentioned positives, followed by financial benefits gained, with some of the experts also specifically mentioning health benefits. Discussion of the positives included the following comments:

> Service to this great nation. These are all volunteers and they are enormously proud of what they are doing. (17: Non-DoD military family expert)

> It is a special person that is a dual citizen, to wear a coat and tie and a uniform. This shows great patriotism and can show patriotism in a small community especially. The reservist is proud to be fighting the GWOT [Global War on Terror]. (11: DoD military family expert)

> Younger service members might actually be making more money when they are activated. Especially if you include tax-free pay and hostile fire pay, they could make more money for a year, which is good for the family. (8: DoD military family expert)

> There are a lot of people that don't have benefits and becoming an active duty family gives them benefits. Although [TRICARE]

can be a pain at times, it is better than nothing if they didn't have benefits to begin with. (2: DoD military family expert)

During the interviews, we asked spouses and service members, "In what ways, if any, has your [spouse's] most recent activation or deployment been a positive experience for your family?" The majority of both spouses and service members—roughly three-fourths of each—provided a positive aspect of the activation or deployment. The positive aspects mentioned most frequently included the following:

- family closeness
- financial gain
- patriotism, pride, or civic responsibility
- independence, confidence, or resilience
- employment and education.

As shown in Figure 5.1, spouses were more likely than were service members to mention the intangible benefits of family closeness; patriotism, pride, or civic responsibility; and independence, confidence, or resilience, but none of the individual positive aspects mentioned were predominant; the most-mentioned positive aspect for spouses, family closeness, was mentioned by only 29 percent of the spouses interviewed. Further, 20 percent of service members and 13 percent of spouses felt that they had not experienced any positives resulting from deployment or activation.

The following sections provide greater detail regarding the positives that were mentioned and the characteristics of individuals who were likely to mention experiencing each positive.

Family Closeness

Twenty percent of service members and 29 percent of spouses mentioned a positive benefit to their family, which we have labeled family closeness. Typical comments included

Figure 5.1
Positives Reported by Service Members and Spouses

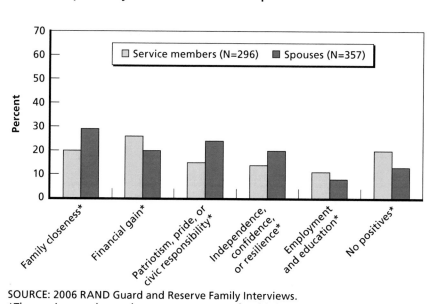

SOURCE: 2006 RAND Guard and Reserve Family Interviews.
*The service member and spouse percentages for this item were statistically different from one another at p<0.10.
RAND *MG645-5.1*

Distance makes the heart grow fonder. It just makes us realize how much we appreciate each other, how important we are in each other's lives. At the end of the day, you realize how important it is to have someone there to talk to when you've had a crappy day, just realize the importance of him being there. (147: Air Force Reserve, E-6's wife)

Because we have such limited and infrequent communication, we are forced to make the most of what we get, and so it has improved our communication skills as husband and wife. And it's improved our marriage because we don't take day-to-day things for granted anymore. (384: Army Reserve, E-6's wife)

The other positive thing would be the way you look at things, a little bit more perspective of what is important, for us it is a strong family nucleus. It really emphasizes that when you are absent. (324: Army National Guard, O-2)

Table 5.1 indicates the characteristics of interviewees most likely to mention this kind of positive aspect. Among service members, those who lived farther from the nearest military installation, those who perceived the notice they received as adequate, and those who had experienced only a single deployment were more likely to note this positive. Among spouses, parents and spouses whose service members had not

Table 5.1
Characteristics Associated with Citing Family Closeness as a Positive Aspect

	Service Members (%)	Spouses (%)
Overall percentage citing family closeness (N=296 service members; N=357 spouses)	20	29
Parental status		
Has children (N=269)		32
No children (N=88)		22
Service member prior active duty		
Yes (N=206)		12
No (N=150)		31
Distance from nearest military installation		
Less than 25 miles (N=120)	14	
25 or more miles (N=176)	24	
Perception of notice adequacy		
Adequate (N=133)	26	
Insufficient (N=55)	13	
Repeat OCONUS deployments		
Yes (N=48)	10	
No (N=248)	22	

SOURCE: 2006 RAND Guard and Reserve Family Interviews.
NOTES: Ns are provided for either service member or spouse, as denoted in the table. When data from both groups are shown, Ns are specified as service member or spouse. All percentages shown are statistically different from one another at p<0.10.

previously served on active duty were more likely than other spouses to mention family closeness as a positive.

Financial Gain

Financial gain was mentioned as a positive aspect of the most recent activation or deployment by 26 percent of the service members and 20 percent of the spouses interviewed. This counters somewhat the common public perception that service members suffer financial setbacks during activation and deployment, but is consistent with the findings of related research. Specifically, Loughran, Klerman, and Martin (2006) found that although 17 percent of reservists studied did experience a loss in earnings, the average change in income during a 2002 or 2003 reserve activation was an increase of roughly $13,500.

Among the interviewees that referred to financial gain, some seemed to share the public perception in their belief that their positive financial experience was unique, as reflected in the following comment: "I think most people lose money when they come on active duty. I actually make more as a reservist" (94: Army Reserve, O-3). Others believed that, even with the additional costs of having a parent absent, they benefited financially: "Even after you subtract the additional costs of child care, we were taking home more money" (551: Marine Forces Reserve, O-3). Other comments included

> I saved up a lot of money while I was over there. The money that I saved kept me and my partner from having to borrow money to start a company. I paid off all of my debt and came back debt-free. (208: Army National Guard, O-3)

> And it helped us to get, we had a little bit more extra income coming in, so we got pretty well established, and we were able to buy a house. And it's been pretty good for us. (587: Army Reserve, E-4's wife)

> Well, we got a pool and my house is remodeled. (66: Army Reserve, E-4)

As Table 5.2 indicates, junior officers, service members who had previously served on active duty, and spouses who lived farther away from the nearest military installation were more likely than their counterparts to mention financial gain.

Patriotism, Pride, or Civic Responsibility

Approximately 15 percent of service members and 24 percent of spouses interviewed mentioned some combination of patriotism, pride, and civic responsibility as a positive. We have grouped these aspects

Table 5.2
Characteristics Associated with Citing Financial Gain as a Positive Aspect

	Service Members (%)	Spouses (%)
Overall percentage citing financial gain (N=296 service members; N=357 spouses)	26	20
Service member pay grade		
E-1 to E-4 (N=69)	23	
E-5 to E-6 (N=146)	23	
O-1 to O-3 (N=81)	33	
Service member prior active duty		
Yes (N=158)	30	
No (N=138)	22	
Distance from nearest military installation		
Less than 100 miles (N=292)		19
100 or more miles (N=31)		35

SOURCE: 2006 RAND Guard and Reserve Family Interviews.

NOTES: Ns are provided for either service member or spouse, as denoted in the table. When data from both groups are shown, Ns are specified as service member or spouse. All percentages shown are statistically different from one another at p<0.10. Shaded cells indicate subsets of the population that are not significantly different from other subsets. For reserve component comparisons among service members, the E-5 to E-6 category differs significantly from the O-1 to O-3 category. The other pay grade comparisons are not significantly different.

together because the comments were so closely related and often mentioned in conjunction with one another. Typical comments discussing patriotism included

> It is always an honor to serve your country. It made me proud, my family is proud. (385: Army National Guard, E-4)

> I believe that it has allowed the children to understand the reasons why we should be proud to be Americans. (47: Army National Guard, E-6's wife)

> The kids and I both look up to him as being a hero. He's doing his call of duty and we are very proud of him. (436: Army National Guard, E-4's wife)

The last two of these comments mentioned children specifically, as one spouse mentioned her children's pride in being American and the other reported that both she and her children were proud of their guardsman. These are closely related to the following comments, which emphasized how the experience taught children of reservists and guardsmen about the responsibility of being an American citizen.

> [I] set a good example for the kids and their friends about serving the country. [I] help[ed] the children see that it is not all about what happens here at home. There is a whole world out there and certain things need to happen for everybody's well being. It was a good experience for everybody. (197: Marine Forces Reserve, E-6)

> Well for one it was something that I felt had to be done. I knew it was going to put a lot of stress on my family. What I showed my children was that yes it was dangerous, it was risky, but there are times in life when you have to make a stand. You have to do what is right. There are things that the U.S. represents to the world and that was what I was doing. (355: Army National Guard, O-3)

Table 5.3 indicates the characteristics of interviewees most likely to mention patriotism, pride, or civic responsibility as a positive aspect

Table 5.3
Characteristics Associated with Citing Patriotism, Pride, or Civic Responsibility as a Positive Aspect

	Service Members (%)	Spouses (%)
Overall percentage citing patriotism, pride, or civic responsibility (N=296 service members; N=357 spouses)	15	24
Age		
25 or less (N=55)		15
26 or more (N=302)		26
Service member pay grade		
E-1 to E-4 (N=69)	9	
E-5 to E-6 (N=146)	21	
O-1 to O-3 (N=81)	10	
Service member reserve component		
Army National Guard (N=104 service members; N=102 spouses)	9	20
Army Reserve (N=74 service members; N=89 spouses)	15	17
Air Force Reserve (N=60 service members; N=83 spouses)	17	28
Marine Forces Reserve (N=58 service members; N=83 spouses)	24	34
Service member prior active duty		
Yes (N=158)	10	
No (N=138)	17	
Perception of notice adequacy		
Adequate (N=161)		30
Insufficient (N=62)		15
Deployment length		
Less than one year (N=180)		28
One year or more (N=169)		21

Table 5.3—Continued

	Service Members (%)	Spouses (%)
Repeat OCONUS deployments		
Yes (N=132)		31
No (N=225)		20

SOURCE: 2006 RAND Guard and Reserve Family Interviews.

NOTES: Ns are provided for either service member or spouse, as denoted in the table. When data from both groups are shown, Ns are specified as service member or spouse. All percentages shown are statistically different from one another at p<0.10. Shaded cells indicate subsets of the population that are not significantly different from other subsets. For pay grade comparisons among service members, the E-5 to E-6 category is significantly different from both the E-1 to E-4 and from O-1 to O-3 categories. The other pay grade comparison is not significantly different. For reserve component comparisons among service members, only the Army National Guard is significantly different from the Marine Forces Reserve. For reserve component comparisons among spouses, the Army Reserve is significantly different from both the Air Force Reserve and from the Marine Forces Reserve. The Army National Guard is also significantly different from the Marine Forces Reserve. The other component comparisons are not significantly different.

of deployment. Mid-grade enlisted personnel, Marine reservists, and those who had not previously served on active duty were more likely to express patriotism, pride, or civic responsibility than others. Among the spouses, older spouses, those married to Marine or Air Force reservists, those who felt they had received adequate notice before deployment, those who had experienced shorter deployments, and those who had experienced multiple deployments were more likely to mention this aspect.

Independence, Confidence, or Resilience

Roughly one-fifth of spouses described a positive change in their independence, confidence, or resilience as a result of the deployment. Almost as many service members also noted this change in their spouse or family. Typical comments included the following:

In a positive way, it made me stronger, and it really did show me that if push comes to shove that I really could deal with managing things on my own. (14: Army National Guard, E-4's wife)

I've become more self-sufficient and learned things that I probably should have known before he left, like our financial status and our bills and just things taking care of the house. (547: Army National Guard, O-2's wife)

It taught my kids more responsibility, helping out around the house with their mom. They have matured, just learned more responsibility while I was gone. My wife is definitely more independent, a lot more self-sufficient; she has learned how to deal with probably more stress than any other wife has had to deal with. She has not intended to be a single parent, but she has learned how to be one. (277: Army National Guard, O-3)

You don't realize how much you rely on your spouse to do things. That is an eye-opener. You learn to do things on your own, which is good. I guess everybody needs to learn to do things on their own and not depend on somebody else. Like learning how to use the weed eater or mowing the grass. (43: Air Force Reserve, E-6's wife)

As Table 5.4 indicates, there were no significant patterns among service members, and there were only three unique characteristics of spouses who were more likely to mention this aspect of deployment. Specifically, only female spouses cited this positive aspect. Also, spouses experiencing their first deployment and those who lived farther from the nearest military installation were more likely to mention this as a positive aspect.

Employment and Education

In addition to asking all interviewees about the positives their families incurred as a result of the activation or deployment, as noted in Chapter Four, we also asked service members about the effect that their ser-

Table 5.4
Characteristics Associated with Citing Independence, Confidence, or Resilience as a Positive Aspect

	Service Members (%)	Spouses (%)
Overall percentage citing independence, confidence, or resilience (N=296 service members; N=357 spouses)	14	20
Gender		
Male (N=12)		0
Female (N=345)		21
Distance from nearest military installation		
Less than 25 miles (N=148)		16
25 or more miles (N=175)		25
Repeat OCONUS deployments		
Yes (N=132)		14
No (N=225)		24

SOURCE: 2006 RAND Guard and Reserve Family Interviews.

NOTES: Ns are provided for either service member or spouse, as denoted in the table. All percentages shown are statistically different from one another at p<0.10.

vice in the National Guard or Reserve had had on their employment and education. While 21 percent of service members discussed some kind of negative effect on their employment and 15 percent of them described a negative effect on education (as discussed in the preceding chapter), a relatively small share, 11 percent of interviewed service members and 8 percent of interviewed spouses, noted a positive effect on their employment or their education.

Generally, the positive benefits to their employment were attributed to the skills or training gained from reserve component service, especially where their reserve or guard occupation related to their civilian work, or to the contacts made during their military service, as shown in the following comments:

> As far as my employment . . . I am a police officer so it has improved my supervisor skills. It has helped with tactical situa-

tions with my employment. I am on an emergency response team and we do a lot of close quarters, urban combat situations, so we have a little team. Especially during train up, that is when we did a lot of training on it, room clearing and taking prisoners. (4: Army National Guard, E-5)

Well, it affected my civilian employment because I was able to get a new job because of what I did in Iraq and what I did in the States during the two years that I was activated. (152: Marine Forces Reserve, O-3)

Those who mentioned positive effects on their education were generally referring to financial support provided by the Reserve Component. These numbers were too small for us to distinguish significant patterns among the respondents.

No Positives

Twenty percent of service members and 13 percent of spouses interviewed expressly stated that their family had experienced no positives stemming from the most recent activation and deployment.[1] As Table 5.5 shows, there were more significant patterns among interviewees with this response. Specifically, interviewed mid-grade enlisted personnel were more likely than junior officers to claim that there were no positive aspects to the deployment. Service members who received less notice and those who described their notice as insufficient were also more likely to say "no positives." Among the spouses interviewed, male spouses, Army Reserve spouses, those who had previously served in the military, and those who described their employers as less than supportive were more likely to claim that there were no positive aspects to the deployment. Additionally, both service members and spouses whose families were less prepared for the deployment were more likely to assert the lack of positives.

---

[1]   An additional one-tenth of service members and an even smaller proportion of spouses skipped this question or otherwise failed to provide an answer.

Table 5.5
Characteristics Associated with Citing No Positives

	Service Members (%)	Spouses (%)
Overall percentage citing no positives (N=296 service members; N=357 spouses)	20	13
Gender		
Male (N=12)		42
Female (N=345)		12
Service member pay grade		
E-1 to E-4 (N=69)	23	
E-5 to E-6 (N=146)	23	
O-1 to O-3 (N=81)	14	
Service member reserve component		
Army National Guard (N=102)		16
Army Reserve (N=89)		21
Air Force Reserve (N=83)		10
Marine Forces Reserve (N=83)		6
Spouse prior military (spouses only)		
Yes (N=34)		24
No (N=323)		12
Employer supportiveness		
Supportive (N=210)		11
Neutral (N=29)		21
Not supportive (N=22)		27
Amount of notice		
One month or less (N=156)	26	
More than one month (N=140)	14	
Perception of notice adequacy		
Adequate (N=133)	17	
Insufficient (N=55)	31	

Table 5.5—Continued

	Service Members (%)	Spouses (%)
Family readiness		
Ready or very ready (N=192 service members; N=214 spouses)	17	11
Somewhat ready (N=43 service members; N=70 spouses)	19	26
Not at all ready (N=51 service members; N=55 spouses)	39	4

SOURCE: 2006 RAND Guard and Reserve Family Interviews.

NOTES: Ns are provided for either service member or spouse, as denoted in the table. When data from both groups are shown, Ns are specified as service member or spouse. All percentages shown are statistically different from one another at p<0.10. Shaded cells indicate subsets of the population that are not significantly different from other subsets. For pay grade comparisons among service members, the E-5 to E-6 category is significantly different from the O-1 to O-3 category. The other pay grade comparisons are not significantly different. For reserve component comparisons among spouses, the Army Reserve is significantly different from both the Air Force Reserve and from the Marine Forces Reserve. The Army National Guard is also significantly different from the Marine Forces Reserve. The other reserve component comparisons are not significantly different.

Discussion

The majority of both spouses and service members mentioned some positive aspects associated with activation and deployment, with a slightly larger proportion of spouses seeing some benefit from the experience. The four benefits mentioned most frequently were family closeness; financial benefit; patriotism, pride, or civic responsibility; and spouse or child independence, confidence, or resilience. Yet, none of these positive aspects were mentioned by more than one-fourth of interview participants. Table 5.6 summarizes the factors we examined in our analysis of the positive aspects and notes those instances in which there is a relationship between a characteristic and the likelihood that a spouse or service member mentioned a particular positive aspect of deployment. Some characteristics, such as age, gender, and parental status, were related to only a few positive aspects. Pay grade, reserve component, and the service member's prior active duty experi-

Table 5.6
Summary of Factors Related to Cited Positives

	Financial Gain	Family Closeness	Patriotism	Independence	No Positives
Individual and situational characteristics					
Age			SP		
Gender				SP	SP
Parental status		SP			SM
Service member pay grade	SM		SM		SM
Service member reserve component			SM, SP		SP
Service member prior active duty	SM	SP	SM		
Spouse prior military (spouses only)					SP
Spouse employer supportiveness (spouses only)					SP
Distance from nearest military installation	SP	SP		SP	
Amount of notice					SM
Perception of notice adequacy		SM	SP		SM
Deployment length			SP		
Repeat OCONUS deployments		SM	SP	SP	
Family readiness					SM, SP

SOURCE: 2006 RAND Guard and Reserve Family Interviews.

NOTES: All relationships listed are statistically significant at p<0.10. SM = Finding present in the service member portion of the sample (N=296); SP = Finding present in the spouse portion of the sample (N=357).

ence were related to several aspects, albeit with different relationships. For example, officers were more likely to mention financial gain, but enlisted personnel were more likely to mention patriotism and to note the lack of positive aspects. Similarly, in those families where the service members had prior active duty experience, the interviewees were more likely to note financial gain but less likely to mention family closeness or patriotism. Experiencing repeat deployments was also associated with mixed outcomes: Those interviewees were less likely to discuss family closeness or spouse independence but more likely to note patriotism. Distance from the nearest military installation had a consistent effect for spouses: Those who lived farther were more likely to note increased financial benefits; independence, confidence, or resilience; and increased family closeness.

CHAPTER SIX

How Well Do Guard and Reserve Families Cope?

Prior research has evaluated the extent to which families cope with deployment, and which types of families report difficulty coping with deployment, but has not offered a precise definition or explanation for coping (see, for example, Caliber Associates, 2003). In this research, we explore the extent to which families share a common understanding of what it means to cope with deployment, which types of families report coping well, and whether there are relationships between reported levels of coping, the problems or positives mentioned by families, and their retention intentions. The relationship between coping and retention intentions is considered later, in Chapter Eight. This chapter focuses on our other research considerations of coping.

Coping is a complex construct; it means different things to different individuals. For some, it may mean successfully enduring stress and hardship; for others, it may just mean surviving. We asked our participants to define what it meant to them. To do so, we employed open-ended questions to discuss family coping. We asked interviewees, "What do you think coping with activation or deployment means for your family?" We also asked interviewees, "How well has your family coped with your recent deployment and why do you say that?" We coded and analyzed the responses to both these questions, and we discuss the results in this chapter.

Defining Coping

We found considerable variation in the interpretation of coping, which supports the premise that coping remains unclearly defined. When asked what coping meant for their family, the majority of respondents—63 percent of service members and 71 percent of spouses—were able to provide a definition of coping. However, the balance of the interviewees were either unable to provide an answer or provided a nonspecific answer, such as "just getting through each day." Those interview participants who did provide a specific definition of coping tended to describe either emotional coping or coping with household responsibilities, which includes child care issues. Figure 6.1 shows the proportions of spouses and service members who provided these two definitions of coping (among those who provided a definition). A relatively small number of participants also mentioned the issue of coping financially, in terms of paying the bills and making ends meet during deployment.

Figure 6.1
Definitions of Coping Provided by Service Members and Spouses

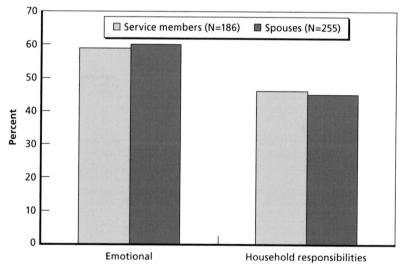

SOURCE: RAND 2006 Guard and Reserve Family Interviews.
RAND MG645-6.1

Emotional Coping

Emotional coping was mentioned most frequently in the definitions provided; 60 percent of spouses and 59 percent of service members who defined coping associated it with emotional well-being. Typical comments from service members and spouses that defined coping as an emotional issue included

> I guess mentally being able to [be] strong when I'm gone. (53: Army National Guard, E-5)

> Means just, best way to put it is just suck it up and get on with the next day. You know you are going to have to accept it. Sitting there, always worrying about it all the time, doesn't help. (197: Marine Forces Reserve, E-6)

> Coping, I would say, is just dealing with the emotional stress of your spouse and parent being gone and dealing with that. (4: Marine Forces Reserve, E-5's wife with two children)

> Not having nervous breakdowns. (83: Army National Guard, E-5's wife)

The common characteristics among the interviewees who tended to provide an emotional definition for coping are displayed in Table 6.1. Younger service members and those without children were more likely to define coping as an emotional issue. Likewise, spouses who were not parents were also more likely to mention this definition, as were those who had been married for less time, those married to junior enlisted personnel, and those who had not previously served in the military. Spouses who reported less notice before the deployment were also more likely to provide an emotional definition for coping.

Coping with Household Responsibilities

Among those who provided a definition for coping, 44 percent of spouses and 45 percent of service members mentioned dealing with the family and household responsibilities as a definition of coping. Typical comments illustrating this definition of coping include the following:

Table 6.1
Characteristics Associated with Defining Coping as an Emotional Issue

	Service Members (%)	Spouses (%)
Age		
25 or less (N=18)	83	
26 or more (N=167)	56	
Marriage length		
2 years or less (N=47)		77
3 years or more (N=208)		56
Parental status		
Has children (N=152 service members; N=195 spouses)	55	57
No children (N=34 service members; N=60 spouses)	74	70
Service member pay grade		
E-1 to E-4 (N=59)		68
E-5 to E-6 (N=121)		62
O-1 to O-3 (N=75)		51
Spouse prior military (spouses only)		
Yes (N=26)		39
No (N=229)		62
Amount of notice		
One month or less (N=113)		68
More than one month (N=136)		53

SOURCE: 2006 RAND Guard and Reserve Family Interviews.

NOTES: Ns are provided for either service member or spouse, as denoted in the table. Ns represent the total number of those who provided a definition for coping. When data from both groups are shown, Ns are specified as service member or spouse. All percentages shown are statistically different from one another at p<0.10. Shaded cells indicate subsets of the population that are not significantly different from other subsets. For pay grade comparisons among spouses, the E-1 to E-4 category is significantly different from the O-1 to O-3 category. The other pay grade comparisons are not significantly different.

Just handling the day to day [and] everything that could, could happen in the home. Taking care of the kids, the house, the pets, the cars. Keeping things running. (8: Army Reserve, E-5's wife with two children)

It means being able to function and carry on as if he were here and take care of everything as if he were here. Make sure everything is taken care of. Making sure all of the bills are paid, the lawn is mowed, the kids are taken care of, the kids have their sporting events, child care issues, my husband is the primary person that picks them up from child care. We didn't have [him here to do that], so I had to make sure that somebody was available to pick the children up from child care. (90: Air Force Reserve, E-5's wife with two children)

It means filling in the gap when I am not here. I mean if I am not here, my wife has to be, she has to be like a mother, she has act like a disciplinarian, she has to do everything that I have to do around the house. (336: Air Force Reserve, E-6 with one child)

Table 6.2 displays the characteristics associated with interviewees who defined coping as a household responsibility issue. Several of the characteristics were common across both service members and spouses. For both groups, older interviewees and those married longer were more likely to define coping this way. Junior enlisted service members and spouses married to junior enlisted personnel were less likely to do so. Army guardsmen and Army reservists were more likely than Marine reservists to define coping as a household responsibility issue. Service members who received less notice were also more likely to define coping as a household responsibility issue.

Among spouses, those who are parents were more likely to define coping as this issue of dealing with matters at home. And, similar to the reserve component pattern among the service members, spouses married to Marine reservists were less likely than the other spouses interviewed to provide a household responsibility definition for coping.

Table 6.2
Characteristics Associated with Defining Coping as a Household Responsibility Issue

	Service Members (%)	Spouses (%)
Age		
25 or less (N=18 service members; N=35 spouses)	11	26
26 or more (N=167 service members; N=220 spouses)	49	47
Marriage length		
2 years or less (N=27 service members; N=47 spouses)	26	28
3 years or more (N=146 service members; N=208 spouses)	49	48
Parental status		
Has children (N=195)		48
No children (N=60)		32
Service member pay grade		
E-1 to E-4 (N=38 service members; N=59 spouses)	34	34
E-5 to E-6 (N=89 service members; N=121 spouses)	54	45
O-1 to O-3 (N=59 service members; N=75 spouses)	54	52
Service member reserve component		
Army National Guard (N=68 service members; N=73 spouses)	49	47
Army Reserve (N=45 service members; N=64 spouses)	56	47
Air Force Reserve (N=35 service members; N=57 spouses)	37	54
Marine Forces Reserve (N=38 service members; N=61 spouses)	32	30

Table 6.2—Continued

	Service Members (%)	Spouses (%)
Amount of notice		
One month or less (N=97)	51	
More than one month (N=89)	38	

SOURCE: 2006 RAND Guard and Reserve Family Interviews.

NOTES: Ns are provided for either service member or spouse, as denoted in the table. Ns represent the total number of those who provided a definition for coping. When data from both groups are shown, Ns are specified as service member or spouse. All percentages shown are statistically different from one another at p<0.10. Shaded cells indicate subsets of the population that are not significantly different from other subsets. For pay grade comparisons among both service members and spouses, the E-1 to E-4 category is significantly different from the O-1 to O-3 category. Other pay grade comparisons are not significantly different. For reserve component comparisons among service members, the Marine Forces Reserve differs significantly from the Army National Guard and the Army Reserve. For reserve component comparisons among the spouses, the Marine Forces Reserve differs significantly from the other three components. Other reserve component comparisons are not significantly different.

How Well Do Families Cope with Deployment?

After inquiring how interview participants defined coping, we also asked a second open-ended question to determine how well they felt their family had coped, or were coping, during deployment. During the coding and analysis, we categorized their responses as indicating that the family was coping or had coped well or very well, moderately well, or poorly. The logic we employed when coding this data was very similar to that discussed in the chapter on readiness (Chapter Three) and used to code the evaluations of how ready families were for deployment. Responses from participants who expressly stated that they coped well, very well, or who did not mention any way in which they had not coped (i.e., implicitly coped well), were coded as "coped well/very well." Typically, those responses were similar to "I think I've been coping very well" (18: Marine Forces Reserve, E-5's wife), whereas responses such as the following were coded as "coped moderately well":

Moderately at best. I'm making it through, but I'm not enjoying it. (66: Army National Guard, E-6's husband)

I think I'm doing okay. There are days I can handle everything that comes my way and days I can't wait to divorce him for putting me through this. (78: Air Force Reserve, E-6's wife)

It ain't always easy, but we're doing pretty good now. (121: Marine Forces Reserve, E-3)

Now, 5 months into it I am coping pretty well. In the beginning I did not do well. [I was] not sleeping, [I was] breaking down into tears. (820: Marine Forces Reserve, O-3's wife)

These comments, along with the others that were categorized as indicative of coping moderately well, often mentioned good days and bad days, or that they had coped differently during different stages of the deployment. Finally, the comments that reflected more serious coping difficulties, from families that had not coped well at all, were included in the final category, "coped poorly." The following are typical comments of those whose responses were coded as "coping poorly":

I didn't cope too well while he was gone. I was depressed and taking depression medication. (73: Army National Guard, E-5's wife)

I had to start seeing a psychologist. Everything that happened was overwhelming. (519: Army National Guard, E-6's wife)

[They coped] very poorly apparently, as I came home to an empty house. She packed up everything I owned and she owned and moved into an apartment. (77: Air Force Reserve, E-6)

We found that in general, the majority of families coped well or very well, as reported by 63 percent of the service members who responded and 62 percent of the spouses. These findings are illustrated in Figure 6.2, which indicates also that spouses were slightly more likely than service members to indicate that their family had done moderately well.

Figure 6.2
Family Coping Levels, as Reported by Service Members and Spouses

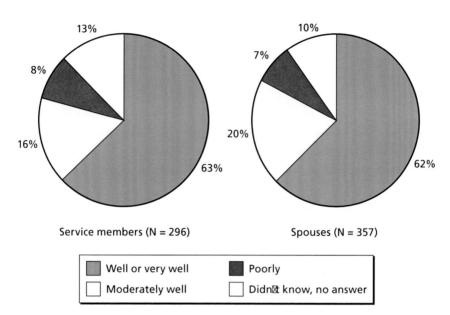

Service members (N = 296)   Spouses (N = 357)

▨	Well or very well	■	Poorly
☐	Moderately well	☐	Didn't know, no answer

SOURCE: 2006 RAND Guard and Reserve Family Interviews.
NOTE: Percentages may not sum to 100 due to rounding.

Which Families Cope Well?

Prior research based on large surveys that could determine the statistical significance of results have indicated that the more senior the pay grade of service members, the longer they had been married, the older their children, and the longer they had lived in their community, the more likely their spouses were to report that they were coping well or very well (Caliber Associates, 2003). Although we did not see many patterns by demographic characteristics of service members, we did see such patterns by spouses, that suggest that—at least among interviewed spouses—our research findings were somewhat consistent with this conventional perspective that more-secure and more-established families tended to cope better. For example, Table 6.3 illustrates that, among our interviewees, spouses who have been married for longer and who are parents tend to cope better, as do spouses of officers and spouses of service members who had previously served on active duty. Additionally, spouses who received more deployment notice were also more likely to report coping well or very well. This had implications by reserve component, given the differences in the amount of notice provided. There are also patterns by distance; spouses who live farther from the drill unit and from the nearest military installation report high coping levels. This may be because these families would not remain in the reserve component, at such large distances, if they did not cope well with deployment.

We also considered the problems and positives that families mentioned and the extent to which those families reported that they were coping well. This additional analysis indicates some relationship between the likelihood that interviewees discussed a particular problem and that they coped well or very well with deployment. Specifically, interviewees who discussed emotional problems, financial and legal problems (service members only), or marital problems were less likely to cope well. Similarly, those who said they had no problems were more likely to say they had coped well or very well. We acknowledge that some respondents may have assessed their coping based on the problems that they were having at the time, or that they recalled having,[1] and thus their perception of coping may be subject to retrospective bias. Nonetheless, there is clearly a relationship between the problems that families experience and their perception of how well they coped.

In general, we did not see the same compelling relationship between the various positive aspects of deployment and the extent to which the family had coped well, with one exception: Spouses who mentioned increased independence and confidence also tended to cope at high levels.

---

[1] Recall that, as noted in Chapter One, while all service members were demobilized, some spouses were married to service members deployed at the time of their interview, and other spouses were married to demobilized service members.

Table 6.3
Characteristics Associated with Family Coping Levels

	Service Members (%)			Spouses (%)		
	Well or Very Well	Moder- ately	Poorly	Well or Very Well	Moder- ately	Poorly
Marriage length						
2 years or less (N=60)				57	27	17
3 years or more (N=257)				72	21	6
Parental status						
Has children (N=240)				73	20	8
No children (N=77)				58	31	10
Service member pay grade						
E-1 to E-4 (N=80)				66	26	8
E-5 to E-6 (N=156)				66	22	12
O-1 to O-3 (N=81)				79	19	3
Service member reserve component						
Army National Guard (N=91)				64	25	11
Army Reserve (N=79)				63	27	10
Air Force Reserve (N=70)				74	24	10
Marine Forces Reserve (N=77)				78	13	9
Service member prior active duty						
Yes (N=182)				73	18	9
No (N=135)				65	28	7
Financial situation						
Comfortable (N=171 SM; N=229 SP)	82	14	5	73	21	6

Table 6.3—Continued

	Service Members (%)			Spouses (%)		
	Well or Very Well	Moder- ately	Poorly	Well or Very Well	Moder- ately	Poorly
Occasional difficulty (N=57 SM; N=70 SP)	61	28	11	57	27	16
Uncomfort- able (N=30 SM; N=17 SP)	37	27	37	71	24	6
Distance from drill unit						
0-99 miles (N=218)				67	23	10
100+ miles (N=70)				77	22	1
Distance from nearest military installation						
Less than 100 miles (N=262)				68	23	9
100 or more miles (N=25)				92	8	0
Amount of notice						
One month or less (N=135)				64	28	8
More than one month (N=171)				74	18	8
Perception of notice adequacy						
Adequate (N=121 SM; N=151 SP)	78	16	7	76	17	7
Insufficient (N=43 SM; N=49 SP)	61	26	14	55	35	10
Family readiness						
Ready or very ready (N=170 SM; N=196 SP)	80	14	7	82	14	4
Somewhat ready (N=39 SM; N=63 SP)	56	36	8	46	40	14

Table 6.3—Continued

	Service Members (%)			Spouses (%)		
	Well or Very Well	Moder-ately	Poorly	Well or Very Well	Moder-ately	Poorly
Not at all ready (N=43 SM; N=42 SP)	54	23	23	50	31	19
Problems and Positives						
Emotional Problems						
Cited problem (N=62 SM; N=121 SP)	60	24	16	57	33	10
Did not cite problem (N=196 SM; N=196 SP)	76	16	8	77	16	7
Financial or Legal Problems						
Cited problem (N=37)	54	27	19			
Did not cite problem (N=221)	75	17	8			
Marital Problems						
Cited problem (N=33 SM; N=34 SP)	39	21	39	50	32	18
Did not cite problem (N=225 SM; N=283 SP)	77	18	5	72	21	7
No Problems						
Cited "no problems" (N=74 SM; N=46 SP)	85	15	0	87	11	2
Did not cite "no problems" (N=184 SM; N=271 SP)	67	20	14	66	24	9

Table 6.3—Continued

	Service Members (%)			Spouses (%)		
	Well or Very Well	Moder- ately	Poorly	Well or Very Well	Moder- ately	Poorly
Independence/Confidence/Resilience Positive						
Cited positive (N=64)				83	16	2
Did not cite positive (N=253)				66	24	10

SOURCE: 2006 RAND Guard and Reserve Family Interviews.

NOTES: SM = service members; SP = spouses. Ns are provided for either service member or spouse, as denoted in the table. Ns represent the total number of those who provided a definition for coping. When data from both groups are shown, Ns are specified as service member or spouse. All percentages shown are statistically different from one another at p<0.10. Shaded cells indicate subsets of the population that are not significantly different from other subsets. For the pay grade comparisons among spouses, the E-5 to E-6 category is significantly different from the O-1 to O-3 category. For reserve component comparisons among spouses, the Air Force Reserve is significantly different from the three other components, and the Army Reserve is significantly different from the Marine Forces Reserve. The other component comparisons are not significantly different.

Discussion

During their interviews, 37 percent of service members and 29 percent of spouses were unable to provide a definition of what coping meant for their family, suggesting that coping is an ambiguous and perhaps confusing concept for families to consider. The definitions that were provided by other respondents varied, but they primarily included emotional coping and coping with day-to-day household responsibilities. More interview participants defined coping as an emotional issue, and there were some patterns reflecting those who were likely to do so, including younger service members and those in new marriages, service members and spouses who were not parents, spouses of junior enlisted personnel, spouses of those who had not previously served on active duty, and spouses who received less deployment notice.

There were more consistent patterns evident among both service members and spouses for those more likely to define coping as a household responsibility issue. These participants appeared to have a more established family life, in that they were more typically older, had been married for longer, and (among spouses) had children. Junior enlisted personnel (and junior enlisted spouses) were less likely to define coping this way, and there were also some differences by component. Additionally, service members who received less notice were more likely to define coping as an issue of juggling the responsibilities of home.

Regardless of the differences in their ability to define coping and in the definitions of coping they actually offered, almost all respondents were able to assess the extent to which their family had coped with deployment. The majority of respondents felt that their family had coped well or very well, with only relatively small portions of respondents indicating that their family had coped poorly. Table 6.4 summarizes the characteristics related to family coping levels. In general, the responses of our interviewed spouses are consistent with the conventional wisdom supported in prior research (such as Caliber Associates, 2003): Spouses in more-established marriages, parents, spouses of junior officers, and spouses of personnel who previously served on active duty tend to report coping well or very well. Also among spouses, those who live farther from drill units and from military installations, and those who received more notice of the deployment, reported higher levels of coping, as did spouses from the Marine Forces Reserve and the Air Force Reserve. There were also some patterns that were consistent for both service members and spouses: Individuals in comfortable financial situations, those who believed they had received adequate notice of the deployment, and those who reported their family as ready for deployment also reported high levels of coping.

Not surprisingly, service members and spouses who reported some of the common problems of deployment, including emotional problems, financial or legal problems (service members only), and marital problems, were less likely to report higher levels of coping. On the other hand, those interviewees who reported no problems or who mentioned increased independence, confidence, or resilience as a positive aspect of deployment (spouses only) tended to report that their family had coped well or very well.

Table 6.4
Summary of Factors Related to Family Coping

	Portion of Interview Sample
Individual and Situational Characteristics	
Marriage length	SP
Parental status	SP
Service member pay grade	SP
Service member reserve component	SP
Service member prior active duty	SP
Financial situation	SM, SP
Distance from drill unit	SP
Distance from nearest military installation	SP
Amount of notice	SP
Perception of notice adequacy	SM, SP
Family readiness	SM, SP
Problems	
Emotional or mental	SM, SP
Financial and legal	SM
Marital	SM, SP
No problems	SP, SM
Positives	
Spouse/child independence	SP

SOURCE: 2006 RAND Guard and Reserve Family Interviews.

NOTES: All relationships listed are statistically significant at $p<0.10$.
SM = Finding present in the service member portion of the sample
(N=296). SP = Finding present in the spouse portion of the sample
(N=357).

We conclude that, despite definitional differences, when asked how their family coped during deployment, most individuals are able to assess roughly how their family had fared during deployment, even though different types of families were challenged by different issues. Further, most guard and reserve families in our study cope well or very well with deployment.

CHAPTER SEVEN

What Resources Do Guard and Reserve Families Use During Deployment?

In addition to considering the problems and positives families experienced as a result of their deployment, we also examined the resources that families turned to for support during deployment. In this chapter, we summarize the key findings related to families' use of both formal or military resources and informal or non-military ones. We also discuss reasons why families may not be accessing these resources, as suggested in both our expert interviews and those with spouses and service members. Finally, we consider issues related to cross-leveling, or otherwise deploying individuals without their usual unit, and how that practice influenced both families' needs for support and their use, broadly speaking, of typical military support resources.

Military and Informal Resources

We asked two separate questions intended to elicit both the formal or military resources and programs used as well as the informal or non-military ones. For brevity, we will refer to them as military and informal resources, respectively. During our interviews with spouses and service members, we described military resources using the following language:

> Military-sponsored family support programs offer services to National Guard/Reserve personnel and their families, particularly during activation and deployments. Such services include

the Family Readiness Group, Military OneSource financial or legal counseling, and assistance with TRICARE.

After inquiring about the family's general awareness and usage of such programs and services during the most recent activation, interviewees were then asked in an open-ended question to specify which ones their family used. After coding and analyzing their responses, we found that a relatively small number of military-sponsored resources were mentioned, and none was cited by a majority of spouses or service members. The most frequently identified military resource was TRICARE, mentioned by just less than half of both spouses and service members (45 percent and 49 percent, respectively). Family support organizations, such as Family Readiness Groups (FRGs) and Key Volunteer Networks, were a close second, cited by 41 percent of spouses and 45 percent of service members. Military OneSource was a distant third, identified by 12 percent of spouses and 10 percent of service members. Figure 7.1 provides a graphical depiction of these responses. Less frequently mentioned resources included financial assistance, legal assistance, and unit or military personnel (distinct from FRG members or Key Volunteers). Spouses and service members tended to discuss all these military programs and services with similar frequency.

Informal resources were assessed somewhat differently in our spouse and service member interviews. Specifically, we asked, "What nonmilitary, informal, or community resources did you [your family] turn to or use during your most recent activation?" Interviewees had a chance to identify, without prompting, the informal resources they relied on. The interviewer then used the following probe: "Informal or civilian resources that have been mentioned to us include those such as extended family, church, and organizations in your community, like the VFW [Veterans of Foreign Wars] or the Red Cross. Can you talk about the extent to which you [your family] used any of these resources?" This probe frequently led to interviewees noting the use of additional resources. The most commonly mentioned informal resource was family, with 57 percent of spouses and 43 percent of service members discussing how they turned to this resource during deployment. About one-third of both groups (36 percent of spouses, 35 percent of

Figure 7.1
Resources Cited by Service Members and Spouses

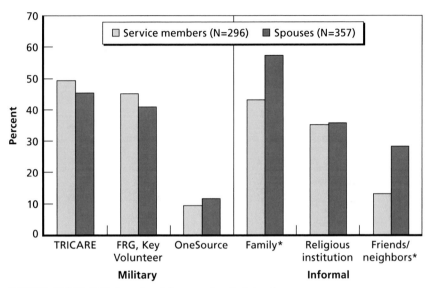

SOURCE: 2006 RAND Guard and Reserve Family Interviews.
*The service member and spouse percentages for this item were statistically different from one another at p<0.10.
RAND MG645-7.1

service members) cited their local religious organization, most typically a church. Twenty-eight percent of spouses and 13 percent of service members referred to friends or neighbors as an informal resource. Response frequencies for these informal resources are provided in Figure 7.1. Additional informal resources that were each mentioned by fewer than 10 percent of spouses and service members include the Red Cross, other military spouses, one's workplace, the Internet, the American Legion, school (either their own or their child's), and the VFW. Unlike military resources, there were some significant differences in how spouses and service members responded to the question; specifically, spouses were more inclined to mention two of the resources, family and friends or neighbors, than were service members.

The spouses and service members we interviewed also differed somewhat in mentioning the use of military or informal resources more broadly. While 82 percent of spouses claimed they used an informal resource of some type, a lower proportion of service members, 72 percent, made a similar assertion about their family. In related findings, service members were more inclined than spouses to report that their family did not use any informal resources at all and almost twice as likely to indicate that their family only used military resources during their deployment. Spouses and service members did resemble one another in how infrequently they asserted that their family did not use *any* resources, either military or informal, during the deployment. Very few spouses and service members—less than 10 percent of them—indicated that their family used neither military nor informal resources.

Not only were there differences between spouses and service members in their responses, as noted above, but there were also patterns among the spouses and among the service members (i.e., within each group) that explain who tended to cite a particular resource. In the sections that follow, we provide additional details on the three most commonly mentioned military and informal resources, including any factors that were associated with the likelihood of using them. The small proportions of interviewees who cited the other resources (less than 10 percent for each) made it difficult to draw inferences about who tended to discuss them. Given that most resources were identified by a limited number of interviewees, we also draw from all our interviews—expert, service member, and spouse—to consider why greater proportions of spouses and service members did not mention using resources more generally.

TRICARE

TRICARE, the most frequently cited military resource, was mentioned with similar frequency in both the spouse and service member portions of the interview sample. While some interviewees simply noted the use of TRICARE with a short answer, others provided additional comments about its value:

> TRICARE is useful because we now have health care insurance. Whereas before, he [her husband, a service member] was unem-

ployed and so, my son and I, my seven-year old and I, were the only ones who had health care insurance because we couldn't afford to get him on there, my spouse, also. Even my oldest child who lives with his father, he [the service member] provides his health care. So, TRICARE has been helpful. (348: Army Reserve, E-6's wife with two children)

TRICARE pretty much for the most part because the kids, they're always sick. Seasons change and the kids get sick and TRICARE helped out a lot. (209: Marine Forces Reserve, E-5 with three children)

TRICARE is good insurance. Once you're in it, it is easy to use, it is not hard to use. And having three kids at home, you didn't ask me about my other two, I actually have five, having three kids at the house, and my wife is a school teacher, so she has insurance through the state, but monetarily, me being on active duty, TRI-CARE saved us money. That was a good thing to have. (22: Army National Guard, O-3 with three children)

We know from the deployment problems discussed in Chapter Four, and from the recommendations for change (discussed in Chapter Nine) that some families had problems with TRICARE, but these comments demonstrate that others viewed it as a useful resource that they appreciated. Both were included in the tally of those who cited TRICARE as a resource. Patterns among spouses and service members who reported using TRICARE are shown Table 7.1. Although, as mentioned above, service members and spouses tended to cite TRI-CARE with similar frequency, a larger number of significant patterns were present in the service member sample. Specifically, older service members (age 26 and up), male service members, and those married longer (three years or more) were more likely to mention TRICARE than were younger service members, female service members, and newlyweds, respectively. For example, 55 percent of service members married at least three years discussed their family's use of TRICARE, compared with 30 percent of newlyweds. Amount of notice was a factor for service members as well; 55 percent of those who received one month's

Table 7.1
Characteristics Associated with Identifying TRICARE as a Resource

	Service Members (%)	Spouses (%)
Age		
25 or less (N=37)	24	
26 or more (N=258)	53	
Gender		
Male (N=270)	51	
Female (N=26)	31	
Marriage length		
2 years or less (N=50)	30	
3 years or more (N=222)	55	
Parental status		
Has children (N=269)		49
No children (N=88)		33
Service member reserve component		
Army National Guard (N=104 service members; N=102 spouses)	64	61
Army Reserve (N=74 service members; N=89 spouses)	42	49
Air Force Reserve (N=60 service members; N=83 spouses)	38	36
Marine Forces Reserve (N=58 service members; N=83 spouses)	43	31
Amount of notice		
One month or less (N=156)	55	
More than one month (N=140)	44	
Deployment length		
Less than one year (N=123 service members; N=180 spouses)	38	41
One year or more (N=173 service members; N=169 spouses)	57	52

SOURCE: 2006 RAND Guard and Reserve Family Interviews.

NOTES: Ns are provided for either service member or spouse, as denoted in the table. When data from both groups are shown, Ns are specified as service member or spouse. All percentages shown are statistically different from one another at $p<0.10$. For reserve component comparisons in the service member group, the Army National Guard is significantly different from those of the other three reserve components. For reserve component comparisons in the spouse group, the Army National Guard and Army Reserve are each significantly different from both the Air Force Reserve and the Marine Forces Reserve. Other reserve component differences are not significantly different.

notice or less discussed using TRICARE, compared with 44 percent of those who received more than one month's notice.

Only one characteristic served as the basis for a spouse-only pattern: parental status. Not surprisingly, spouses with children were more likely to cite TRICARE as a resource than those without children (49 percent versus 33 percent). The final two attributes on which patterns were based, reserve component and deployment length, were significant factors for both service members and spouses. Army guardsmen were more likely to report using TRICARE than were reservists from the three other components included in our study, and spouses married to Army guardsmen or to Army reservists were also more likely to do so than were those married to either Air Force reservists or Marine reservists. In addition, both service members and spouses that experienced a deployment of one year or longer were more inclined to use TRICARE. Fifty-seven percent of service members and 52 percent of spouses who reported a deployment of one year or longer did so, compared with 38 percent of service members and 41 percent of spouses who reported a deployment of less than one year. This may be due in part to civilian employer benefits coverage ceasing after a period of time.

Family Support Organizations

A military resource of a different, nonpecuniary nature, family support organizations, such as FRGs and Key Volunteer Networks, was the second most-cited military resource for both the spouses and the service members we interviewed. Representative comments about this type of military resource included

> Family Readiness Group was good. I had a very good Family Readiness Group Coordinator, and she was very informative. She was giving out a lot of information to all of the spouses in our unit. (204: Army National Guard, female O-3)

> [T]he FRG leader has been very helpful though too. Whenever there is meeting, she usually emails or calls to let us know there's going to be a meeting or any special things for the kids going on. She'll usually let us know that's going on. She also put together a

wives retreat, so that all the wives could get together and converse, and just meet one another and spend time with one another, and relate with one another as far as things that are going on. (23: Army National Guard, E-4's wife)

Living away from the base as far away as we do, I still feel very isolated even though we have the Family Readiness Group. I don't feel like they have been doing a lot to help other than giving me information about the guys. Supposedly the goal of the program is to make sure everybody is OK while they are gone and I don't feel like they have done that well enough. (149: Marine Forces Reserve, E-5's wife)

I just turn to our key volunteer program with other wives. We needed a network. We were able to communicate, we were able to always know what's going on with our husbands. And that really helped. (151: Marine Forces Reserve, O-2's wife)

As the comments above suggest, family support organizations were discussed in both favorable and somewhat unfavorable terms, and thus some of the suggestions for change discussed in Chapter Nine address these organizations. Table 7.2 provides a breakdown of significant response patterns for service members and spouses. There were several common findings; for both service members and spouses, we observed patterns based on pay grade, reserve component, amount of notice, and deployment length. For example, junior enlisted personnel and spouses of junior enlisted personnel were more likely to mention turning to a family support group during deployment. In addition, as with TRICARE, those affiliated with the Army National Guard tended to differ from other interviewees in their tendency to cite family support groups. Sixty-five percent of both Army guardsmen and spouses married to Army guardsmen reported using family support organizations, proportions significantly greater than comparable ones from the other three reserve components. Further, among service members, Air Force reservists were considerably less likely to mention using a family support organization, and, among spouses, those married to Army reservists cited a family support organization as a resource more

Table 7.2
Characteristics Associated with Identifying a Family Support Organization as a Resource

	Service Members (%)	Spouses (%)
Gender		
Male (N=270)	47	
Female (N=26)	27	
College degree		
Yes (N=125)	38	
No (N=171)	50	
Service member pay grade		
E-1 to E-4 (N=69 service members; N=90 spouses)	57	49
E-5 to E-6 (N=146 service members; N=174 spouses)	43	39
O-1 to O-3 (N=81 service members; N=93 spouses)	41	37
Service member reserve component		
Army National Guard (N=104 service members; N=102 spouses)	65	65
Army Reserve (N=74 service members; N=89 spouses)	42	45
Air Force Reserve (N=60 service members; N=83 spouses)	20	19
Marine Forces Reserve (N=58 service members; N=83 spouses)	40	29
Service member prior active duty		
Yes (N=206)		35
No (N=151)		49
Distance from drill unit		
Less than 100 miles (N=245)		47
100 or more miles (N=77)		27
Amount of notice		
One month or less (N=157)		36
More than one month (N=189)		46

Table 7.2—Continued

	Service Members (%)	Spouses (%)
Deployment length		
Less than one year (N=123 service members; N=180 spouses)	29	32
One year or more (N=173 service members; N=169 spouses)	57	52
Family readiness		
Ready or very ready (N=192)	48	
Somewhat ready (N=43)	49	
Not at all ready (N=51)	29	

SOURCE: 2006 RAND Guard and Reserve Family Interviews.

NOTES: Ns are provided for either service member or spouse, as denoted in the table. When data from both groups are shown, Ns are specified as service member or spouse. All percentages shown are statistically different from one another at p<0.10. Shading indicates a subset of population that is not significantly different from other subsets. For pay grade comparisons in the service member group, the E-1 to E-4 category is significantly different from the E-5 to E-6 and O-1 to O-3 categories. The other pay grade comparison is not significantly different. For reserve component comparisons in the service member group, the Army National Guard is significantly different from that of the three other components, and the Air Force Reserve is also statistically different from the Army Reserve and the Marine Forces Reserve. For reserve component comparisons in the spouse group, the Army National Guard is significantly different from the three other components, and the Army Reserve is also statistically different from the Air Force Reserve and the Marine Forces Reserve. Other reserve component comparisons are not significantly different.

frequently than did those married to Air Force reservists or Marine reservists. Lastly, both spouses and service members who experienced a longer deployment were more likely to discuss their use of a family support organization; 57 percent of service members deployed one year or longer and 52 percent of spouses married to a service member deployed one year or longer did so, compared with 29 percent of service members and 32 percent of spouses who experienced a deployment under one year in length.

There were also several patterns evident only among service members or only among spouses. In the service member sample, we found that male service members and those without a college degree tended to discuss their family's use of a family support organization more often than female service members and those with a college degree, respectively. The gender-related finding is consistent with an aforementioned observation (in Chapter Two) by a military family expert, who noted that husbands of service members often do not feel connected to family readiness services available to spouses. In addition, there was a relationship between family readiness and mention of a family support organization: Service members who indicated that their family was not at all ready for deployment were less likely to mention family support organizations as a resource than service members who reported their family as being ready/very ready or somewhat ready. Since this was a cross-sectional study, however, we do not know whether families tended to have a higher level of readiness because they turned to family support organizations or families already at a high level of readiness knew to rely on a family support organization as a resource.

Lastly, three distinct patterns were apparent within the spouse sample. Those married to service members with prior active duty experience were less likely to mention using family support organizations than were those married to service members without such experience. Perhaps as a result of their greater experience base, these spouses are more knowledgeable and did not perceive a need for family support organizations. Amount of notice also accounted for differences in spouse responses: Those who received more notice also tended to mention family support groups more often. In addition, spouses living closer to the service member's drill unit were more likely to cite a family support organization as a resource. Forty-seven percent of spouses living within 100 miles of the drill unit did so, compared with just 27 percent of those living 100 or more miles away. This suggests that the distance and time required to travel to a family support organization meeting may have been a factor for the spouses in our study.

Military OneSource

Military OneSource, cited by 11 percent of service members and spouses overall, was third in terms of frequency of mention. Comments about Military OneSource tended to refer to its ability to provide information on a wide array of topics:

Military OneSource. You can always go there online and call and they give you the right amount of information you need. (145: Army Reserve, E-5)

Oh my goodness. Military OneSource. It answers a lot of questions regarding child care, summer programs, and as far as school, monies for school, like scholarships and such. And they were quick, and they were very patient with my questions. (58: Marine Forces Reserve, E-5's wife with two children)

Military OneSource, very useful. When I called them, having trouble with the kids, they were able to direct me to a family counselor that we were able to go to and who actually offered us six sessions that were greatly needed by us. I would have never known about that or anything without them. (653: Army National Guard, E-4's wife with two children)

Attributes related to the frequency with which Military One-Source was mentioned are listed in Table 7.3. Service member pay grade, service member reserve component, current family financial situation, and deployment length all were sources of significant differences in terms of how frequently service members and spouses cited Military OneSource. Among service members, junior officers were more likely to mention Military OneSource than were mid-grade enlisted personnel, and, among spouses, those married to junior enlisted personnel were more likely to do so than were those married to either mid-grade enlisted personnel or junior officers. With respect to reserve component, among service members, Marine reservists were less likely to report using Military OneSource than were Army guardsmen, and, among spouses, those married to Air Force reservists were far less likely that those affiliated with other reserve components to mention Military OneSource.

While patterns based on pay grade and reserve component manifested differently for service members and spouses, those stemming from family finances and deployment lengths were similar for both groups. For service members and spouses, those with uncomfortable family finances at the time of their interview tended to cite Military

Table 7.3
Characteristics Associated with Identifying Military OneSource as a Resource

	Service Members (%)	Spouses (%)
Marriage length		
2 years or less (N=71)		4
3 years or more (N=286)		14
Parental status		
Has children (N=269)		14
No children (N=88)		6
Service member pay grade		
E-1 to E-4 (N=69 service members; N=90 spouses)	12	20
E-5 to E-6 (N=146 service members; N=174 spouses)	6	9
O-1 to O-3 (N=81 service members; N=93 spouses)	14	9
Service member reserve component		
Army National Guard (N=104 service members; N=102 spouses)	15	16
Army Reserve (N=74 service members; N=89 spouses)	8	16
Air Force Reserve (N=60 service members; N=83 spouses)	7	1
Marine Forces Reserve (N=58 service members; N=83 spouses)	3	13
Service member prior active duty		
Yes (N=206)		9
No (N=151)		15
Financial situation		
Comfortable	8	10
Occasional difficulty	8	12
Uncomfortable	20	27

Table 7.3—Continued

	Service Members (%)	Spouses (%)
Perception of notice adequacy		
Adequate (N=161)		13
Insufficient (N=62)		24
Deployment length		
Less than one year (N=123 service members; N=180 spouses)	6	8
One year or more (N=173 service members; N=169 spouses)	12	16
Family readiness		
Ready or very ready (N=214)		9
Somewhat ready (N=70)		14
Not at all ready (N=55)		24

SOURCE: 2006 RAND Guard and Reserve Family Interviews.

NOTES: Ns are provided for either service member or spouse, as denoted in the table. When data from both groups are shown, Ns are specified as service member or spouse. All percentages shown are statistically different from one another at p<0.10. Shading indicates a subset of population that is not significantly different from other subsets. For pay grade comparisons in the spouse group, the E-1 to E-4 category is significantly different from the E-5 to E-6 and O-1 to O-3 categories. The other pay grade comparison is not significantly different. For reserve component comparisons in the spouse group, the Air Force Reserve is significantly different from the three other components. Other reserve component comparisons are not significantly different.

OneSource. Twenty-seven percent of spouses and 20 percent of service members in direr financial straits discussed Military OneSource, compared with 8 to 12 percent of interviewees with a more positive financial outlook (proportions varied from within this range depending on the type of interviewee and financial category). In addition, for both groups, twice as many interviewees who experienced a deployment of one year or more identified Military OneSource as a resource. Twelve percent of service members and 16 percent of spouses who experienced a longer deployment did so, versus just 6 percent of service members and 8 percent of spouses who reported a deployment length of less than one year.

Although there were no response patterns present among only the service members we interviewed, five characteristics accounted for significant differences among spouses with respect to Military One-Source: marriage length, parental status, service member prior active duty experience, perceptions of notice adequacy, and family readiness. Newlyweds and those without children were less likely to mention Military OneSource than spouses married longer and those with children, respectively. Conversely, those married to service members without prior active duty experience discussed using Military One-Source more frequently than spouses whose service members served in the active duty military prior to joining the Guard or Reserve. Perceptions of notice adequacy and family readiness also served as a basis for differences; spouses who felt the notice received was insufficient and those who characterized their family as not at all ready for deployment both tended to rely on Military OneSource. Twenty-four percent of spouses who reported insufficient notice also mentioned Military One-Source, compared with 13 percent of spouses who deemed their notice adequate. Similarly, 24 percent of spouses who felt that their family was not ready cited Military OneSource, compared with 14 percent of those whose families were somewhat ready and just 9 percent of those whose families were ready or very ready.

Family

Turning our attention to the informal military resources relied on by our interviewees, family was the top choice of both service members and spouses. As mentioned earlier, however, spouses were even more inclined to cite help from family than were service members, and this difference, listed in the second row of Table 7.4, was statistically significant. It is possible that service members were less familiar with—or maybe even unaware of—their family's use of informal resources, such as extended family. Comments about the role of extended family (e.g., parents or siblings of the spouse or service member) in supporting guard and reserve families, such as the ones below, suggest that they filled a variety of needs:

Table 7.4
Characteristics Associated with Identifying Extended Family as a Resource

	Service Members (%)	Spouses (%)
Overall percentage identifying extended family (N=296 service members; N=357 spouses)	43	57
College degree		
Yes (N=195)		65
No (N=162)		49
Service member pay grade		
E-1 to E-4 (N=90)		52
E-5 to E-6 (N=174)		54
O-1 to O-3 (N=93)		69
Service member reserve component		
Army National Guard (N=102)		55
Army Reserve (N=89)		49
Air Force Reserve (N=83)		58
Marine Forces Reserve (N=83)		69
Service member prior active duty		
Yes (N=158)	39	
No (N=138)	49	
Spouse prior military (spouses only)		
Yes (N=34)		41
No (N=323)		59
Financial situation		
Comfortable (N=255)		62
Occasional difficulty (N=74)		46
Uncomfortable (N=26)		50
Distance from nearest military installation		
Less than 100 miles (N=292)		58
100 or more miles (N=31)		39

Table 7.4—Continued

	Service Members (%)	Spouses (%)
Amount of notice		
One month or less (N=157)		53
More than one month (N=189)		62
Perception of notice adequacy		
Adequate (N=133)	46	
Insufficient (N=55)	33	

SOURCE: 2006 RAND Guard and Reserve Family Interviews.

NOTES: Ns are provided for either service member or spouse, as denoted in the table. When data from both groups are shown, Ns are specified as service member or spouse. All percentages shown are statistically different from one another at p<0.10. Shading indicates a subset of population that is not significantly different from other subsets. For pay grade comparisons in the spouse group, the O-1 to O-3 category is significantly different from the E-1 to E-4 and E-5 to E-6 categories. The other pay grade comparison is not significantly different. For reserve component comparisons in the spouse group, the Marine Forces Reserve is significantly different from Army National Guard and the Army Reserve. The other reserve component comparisons are not significantly different.

> Extended family. They are my rock. They help me through everything. (124: Marine Forces Reserve, E-5's wife)

> Mainly I just used the extended family and that was more for stuff like babysitting and helping out with things around the house I can't fix myself. (370: Air Force Reserve, E-4's wife with four children)

> We have a group of family members that are available for my wife when I'm deployed. Pretty much it means that my wife has a group of family members that are close to us. Example is my mother or my father who are available to her at her beck and call. If something happens to me or she needs somebody to talk to, my family's there for her. (16: Army National Guard, E-5)

Table 7.4 also features a summary of who tended to cite extended family as a resource more frequently among our interviewees. In contrast to the military resources already discussed, for extended family there were no common patterns across both the service member and spouse samples. Among the service members that we interviewed, those who had prior active duty experience and those who felt the notice received was insufficient were less inclined to cite extended family as a resource than were service members lacking prior active duty experience and those who deemed their notice adequate, respectively.

More patterns were evident among the interviewed spouses. Those with a college degree, those currently with comfortable family finances, and those living within 100 miles of the nearest military installation were all more likely to cite extended family as a resource. Spouses married to junior officers and those married to Marine reservists also were more inclined to discuss how their family helped them during deployment than were spouses married to service members in other ranks and those married to either Army guardsmen or Army reservists. On the other hand, spouses who had prior military experience were less likely to mention turning to extended family for support: 41 percent of spouses with prior military experience identified this type of informal resource, compared with 59 percent of those lacking it. Lastly, the amount of notice received was a basis for response differences among the spouses: Those who received more than one month's notice were more likely to mention extended family in their discussion of support resources.

Religious Organizations

Religious organizations, primarily churches, were the second most commonly cited informal resources for both the service members and spouses in our study. While, as one would expect, religious organizations were a source of spiritual and emotional sustenance, the comments that follow show they also provided different types of support:

> The church was great, they helped a lot. They called on the wife quite often, asked her is she needed anything or anything like that. That was great. Sent me packages over there. That was great. Some of our church members are former military and they were deployed during Desert Storm and they understood, so it was really great. They could give my wife advice on how things

worked, things like that over there. (148: Army National Guard, E-6)

I would probably say church because they would give you faith in addition to support in any facet. If I had to move, they were there to help. (292: Marine Forces Reserve, O-3's wife)

Probably my church because a lot of my family was not in this immediate area, and my church was basically where I was pretty much at home. These people knew me, they knew my husband, they knew what the situation was. I think in the long run, not only did they give emotional support but there were a lot of times here when I just needed a break. So I would say that the church was the biggest contributor of help whenever I needed it. (331: Army National Guard, E-6's wife)

As Table 7.5 shows, the church and other religious organizations were especially a resource for those experiencing longer deployments: 42 percent of service members and 41 percent of spouses with a deployment of one year or longer identified a religious institution as a resource, compared with 26 percent of service members and 31 percent of spouses with a shorter deployment. In addition, distance was a common factor for both spouses and service members: Those living farther from the drill unit were more likely to mention their family's turning to church or another religious organization during deployment, as were those living farther from the nearest military installation. The largest such difference was present in the spouse sample: 41 percent of spouses living 25 miles away or more from their service member's drill unit mentioned a religious organization as a resource, compared with 27 percent of spouses living less than 25 miles away from the drill unit.

Other findings present in the service member portion of the interview sample pertain to parental status, college degree, and reserve component. Specifically, service members with children and those lacking a college degree were less likely to discuss their family's reliance on a religious organization during their deployment. Conversely, Army guardsmen were more inclined to cite a religious organization than were other

Table 7.5
Characteristics Associated with Identifying a Religious Institution as a Resource

	Service Members (%)	Spouses (%)
Age		
25 or less (N=55)		26
26 or more (N=302)		38
Marriage length		
2 years or less (N=71)		27
3 years or more (N=286)		38
Parental status		
Has children (N=232)	32	
No children (N=64)	48	
College degree		
Yes (N=125)	42	
No (N=171)	30	
Service member pay grade		
E-1 to E-4 (N=90)		38
E-5 to E-6 (N=174)		32
O-1 to O-3 (N=93)		42
Service member reserve component		
Army National Guard (N=104)	48	
Army Reserve (N=74)	31	
Air Force Reserve (N=60)	32	
Marine Forces Reserve (N=58)	21	
Distance from drill unit		
Less than 25 miles (N=83 service members; N=94 spouses)	28	27
25 or more miles (N=213 service members; N=228 spouses)	38	41
Distance from nearest military installation		
Less than 25 miles (N=210 service members; N=148 spouses)	29	30
25 or more miles (N=176 service members; N=175 spouses)	39	40

Table 7.5—Continued

	Service Members (%)	Spouses (%)
Deployment length		
Less than one year (N=123 service members; N=180 spouses)	26	31
One year or more (N=173 service members; N=169 spouses)	42	41

SOURCE: 2006 RAND Guard and Reserve Family Interviews.

NOTES: Ns are provided for either service member or spouse, as denoted in the table. When data from both groups are shown, Ns are specified as service member or spouse. All percentages shown are statistically different from one another at p<0.10. Shading indicates a subset of population that is not significantly different from other subsets. For reserve component comparisons in the spouse group, the Army National Guard is significantly different from the other components. Other reserve component comparisons are not significantly different.

reserve personnel: 48 percent of Army guardsmen mentioned church or another type of religious institution, compared with 21 percent to 32 percent of service members from the other three components included in our study.

The spouse sample also featured three unique patterns with respect to religious organizations: those related to age, marriage length, and service member pay grade. More mature spouses, as suggested by age and marriage length, were more likely to identify a religious organization as a resource during their service member's deployment. Thirty-eight percent of both spouses age 26 or older and spouses married at least three years mentioned this type of informal resource, while only 26 percent of those age 25 or younger and 27 percent of newlyweds expressed a similar sentiment. With respect to pay grade, junior officers' spouses were more likely than spouses of mid-grade enlisted personnel to have discussed turning to a religious organization for support.

Friends and Neighbors

Twenty-eight percent of spouses and 13 percent of service members cited friends and neighbors, making it the third most frequently discussed informal resource for both groups. This difference in proportions, displayed in Table 7.6, was statistically significant. Like extended

Table 7.6
Characteristics Associated with Identifying Friends and Neighbors as a Resource

	Service Members (%)	Spouses (%)
Overall percentage identifying friends and neighbors (N=296 service members; N=357 spouses)	13	28
Age		
25 or less (N=55)		18
26 or more (N=302)		30
Marriage length		
2 years or less (N=71)		20
3 years or more (N=286)		30
College degree		
Yes (N=125 service members; N=195 spouses)	22	36
No (N=171 service members; N=162 spouses)	7	19
Service member pay grade		
E-1 to E-4 (N=69 service members; N=90 spouses)	10	22
E-5 to E-6 (N=146 service members; N=174 spouses)	10	25
O-1 to O-3 (N=81 service members; N=93 spouses)	22	41
Service member reserve component		
Army National Guard (N=102)		28
Army Reserve (N=89)		21
Air Force Reserve (N=83)		34
Marine Forces Reserve (N=83)		30
Financial situation		
Comfortable (N=255)		29
Occasional difficulty (N=74)		20
Uncomfortable (N=26)		42

Table 7.6—Continued

	Service Members (%)	Spouses (%)
Perception of notice adequacy		
Adequate (N=133)	6	
Insufficient (N=55)	17	
Family readiness		
Ready or very ready (N=214)		29
Somewhat ready (N=70)		37
Not at all ready (N=55)		15

SOURCES: 2006 RAND Guard and Reserve Family Interviews.

NOTES: Ns are provided for either service member or spouse, as denoted in the table. When data from both groups are shown, Ns are specified as service member or spouse. All percentages shown are statistically different from one another at p<0.10. Shading indicates a subset of population that is not significantly different from other subsets. For pay grade comparisons in both the service member and spouse groups, the O-1 to O-3 category is significantly different from the E-1 to E-4 and E-5 to E-6 categories. The other pay grade comparison is not significantly different.

family and religious organizations, friends and neighbors supported families in a number of ways. The comments that follow provide some examples:

> [M]y friends, because when I'm having one of those days when no matter what you do, you just feel like you can't go on, you can pick up the phone and you can call them anytime, and they're friends, and they love you, and they accept you, and they tell you they understand, and tell you it's okay, and tell you, what do you need me to do, and I'll be there for you, and that's what you need to keep going. (165: Marine Forces Reserve, E-4's wife)

> My friends. I think my friends have been the best because my husband's base is so far away, that I haven't talked to many of them, plus I didn't know them very well, because he's only usually a reservist, so he only goes down like once a month, so I don't know them very often. But I have called my friends a lot for help on things, or his friends. They seem to be the ones that act the

quickest, so I use them the most. (174: Air Force Reserve, E-4's wife)

You become very heavily dependent on friends and neighbors all of a sudden because the husband or even the wife isn't there to do those kinds of motherly or in this case husbandly duties. You rely on friends and neighbors. In this case, my wife did, enormously, and it worked out. We have great friends and neighbors, I guess. (277: Army National Guard, O-3)

Table 7.6 also features response patterns present among service members, spouses, or both groups with respect to their mention of friends and neighbors as an informal resource. Common findings include those related to college degree and service member pay grade. Both service members and spouses with a college degree were more inclined to cite friends and neighbors as an informal resource than were their less educated counterparts. In addition, both junior officers and spouses of junior officers were more likely to do so than were enlisted personnel and spouses of enlisted personnel. For example, 41 percent of junior officer spouses identified friends or neighbors as a source of support, compared with 25 percent of those married to mid-grade enlisted personnel and 22 percent of those married to junior enlisted personnel.

Turning our attention toward patterns evident only among service members or spouses, perceptions of notice adequacy were related to service members' tendency to discuss friends and neighbors as a resource. Seventeen percent of service members who felt their notice was insufficient also mentioned the support of their friends and neighbors, while only 6 percent of service members who described their notice as adequate did so. The remaining patterns pertained to the spouse portion of the interview sample. As with religious organizations, spouses who were more mature, as suggested by age and marriage length, were more likely to mention friends and neighbors as a resource. Thirty percent of spouses age 26 or older and the same percentage of spouses married at least three years discussed the support provided by their friends and neighbors, compared with 18 percent of younger spouses and 20 percent of newlyweds. Spouses with a college degree and those reporting uncomfortable family finances were more likely to do so as well. One comparison based on reserve component was also apparent: Spouses married to Air Force reservists tended to mention their friends and neighbors significantly more frequently than did spouses of Army reservists. Finally, perceptions of family readiness were related to how frequently friends and neighbors were cited as a resource. Spouses who felt that their family was not at all ready for deployment were less likely to mention friends and neighbors than were those with more favorable perceptions of their family's readiness. Only 15 percent of spouses whose families were not ready at all spoke about the support of friends and neighbors, compared with 37 percent of spouses whose families were somewhat ready and 29 percent of those whose families were ready or very ready.

Possible Explanations for Limited Use of Resources

Although the vast majority of the spouses and service members we interviewed used some type of military or informal resource, no single resource other than family was mentioned by a majority of spouses or service members. In this section, we consider potential reasons for this outcome, as suggested by the military family experts, spouses, and service members we interviewed.

Explanations for the Limited Use of Military Programs or Resources

During their interviews, spouses[1] who indicated they were aware of yet did not use any military programs or services were asked in a follow up question to explain why. In total, 116 of the 357 spouses interviewed indicated that they had not used military programs or services. Of those, 55 percent of them indicated they and their family did not use military resources, such as TRICARE and family support organizations, because they did not need them. Such comments included

---

[1]   We did not pose this question to the service member portion of the sample because we were not confident that service members would be able to explain why their spouse opted not to use a particular service or program.

> I have my own health insurance and do not have to rely on TRI-CARE during this time. (239: Marine Forces Reserve, O-3's wife)

> I have over twenty years service in the military and I have a lot of experience with deployment. (202: Marine Forces Reserve, O-3's husband)

> I have family and friend support, so I use them instead. (128: Army Reserve, E-4's wife)

The last comment suggests that some spouses may have turned to informal resources in lieu of military programs and services, but we did not expressly ask about this during the interview and consequently cannot comment on how widespread this substitution is.

The second most frequently provided reason pertained to accessibility and was offered by 30 percent of the spouses who had not used military resources. Their remarks pertain to a lack of accessibility typically stemming from distance, time, or child-related constraints, as follows:

> I have to go far—50 miles—to where the base is. (102: Marine Forces Reserve, E-3's wife)

> Like right now I work Monday through Friday, from like 8 to 5, and by the time I get home, it's late. And Saturday I normally take my daughter to Chinese school and then piano lessons and art, and I only have Sundays left. That's why I haven't been using a lot of military stuff. (344: Army National Guard, O-3's wife with one child)

Less than 10 percent of spouses who were asked this question stated that they did not use military resources because they did not view them to be worthwhile or of sufficient quality. Their comments included:

> I wasn't impressed. The first time I had a bad experience. I had a guy call twice during 67 days to ask how I was doing. I have

friends and family that are more up on top of things. (43: Air Force Reserve, E-6's wife)

TRICARE, when I listened to that program, it seemed like it was more of a hassle than my own insurance. (87: Army National Guard, E-4's wife)

While the proportion of spouses who did not use military services for this reason is small, similar views were expressed by other spouses, as well as by service members, who *did* access deployment-related military resources. Their comments were considered in Chapter Four, in the context of our analysis of families' problems, and will also be covered in Chapter Nine when we discuss service members' and spouses' suggestions for improvement.

A comparable number of spouses indicated they did not use military resources because they were not knowledgeable about them. The small number of comments included the following:

I would have done TRICARE if I had known about it, but I didn't. I didn't know what kind of services they provide. (147: Air Force Reserve, E-6's wife)

I didn't ever know when Family Readiness Group meetings were. I never heard of OneSource, so I couldn't use it. I didn't know what it is. (641: Army National Guard, E-4's wife)

Because I didn't know who to go to and they did not tell me who to contact. (168: Army Reserve, O-3's wife)

Explanations for the Limited Use of Informal Programs or Resources

Although we did not ask spouses or service members a similar explicit question about why they (or their family) did not use any informal resources, some of them provided explanations without prompting and thus offer useful insights. As with military resources, there were individuals who felt they had no need for informal resources as well as those who were unaware of informal or nonmilitary resources available

to them. This second point especially pertained to the nonprofit organizations we mentioned in our standard follow-up probe, as follows:

> I don't think we used the Red Cross or the VFW. We weren't
> aware of some of those venues [listed in the standard probe]
> because one, I wasn't a veteran, and so I thought that was one
> of the requirements for VFW. But we have recently joined the
> American Legion, that organization, but we weren't even aware
> of that until we came back so we weren't aware of those type of
> sources. (123: Army Reserve, O-3)

> As far as the VFW or the [American] Legion or anything like
> that, I really haven't used or heard of anything in our area. (765:
> Army National Guard, E-4's spouse)

The experts we interviewed also echoed several of these themes in
their discussion of organizational challenges to supporting guard and
reserve families. For example, they acknowledged that some families
are simply unaware their programs and services exist. As one expert
from a military advocacy organization noted, "Many families don't
realize that organizations like ours exist to help them. It is important
for them to understand that we are here to help them" (22: Non-DoD
military family expert). Experts identified two main problems that sty-
mied their efforts to make guard and reserve families more aware of
the services their organizations offered: geographic dispersal and not
knowing how or where to reach them. They explained that not only are
many families far away from a military installation, but they are scattered, so to speak, throughout the country:

> Expert: Geographic dispersal is our biggest challenge.
> RAND: Really, even with your hundreds of family centers?
> Expert: Yes, because it's still a challenge to ensure that the net-
> work connects continually. And we may not have the right kind
> of local experts, like TRICARE experts, at each location. (7:
> DoD military family expert)

> We have a difficult time reaching out to our families because they
> are so dispersed geographically. Our regional readiness staff usu-

ally consists of a group of three to five people who cover five to eight states. That makes it difficult to reach out to all of these people. They are all over the place and it is difficult to get them into meetings or to the mobilization briefings, and to execute coordinated efforts. We have to mail stuff out. (15: DoD military family expert)

While the second comment suggests one way to counter geographic dispersion, other remarks from experts suggest that sometimes they are unable to communicate with guard and reserve families via mail or other remote means. Their comments about this dilemma[2] included:

Having the correct contact information for families [is a challenge]. Some soldiers don't see the need in giving us current information for their families, because they consider themselves stable families. Also, our mailings might appear to be junk mail and I'm not sure that everyone is reading the material we send them. But we also try to call, which is a problem if no home phone exists, [and they only have] cell phones instead. And we email, but again this is not a guarantee. (13: DoD military family expert)

Reaching them—that is the biggest challenge. We try to reach out but it's hard to find families sometimes. We pass out cards to the mobilized member so they can provide a mailing address and we will send an information package to their family. But if the sailor doesn't want or care about their families getting an information package, then we don't get an address. (12: DoD military family expert)

An additional challenge that was very salient to the experts in our study but less so to the service members and spouses we interviewed was related to insufficient resources. Both DoD professionals and those affiliated with either nonprofit service providers or advocacy groups

---

[2]   This study's experience corroborates the views expressed during expert interviews: As discussed in Chapter One, our own data collection efforts were also impeded by difficulties in reaching guard and reserve families.

discussed how their efforts to help families were limited by either a lack of funding or staff, as follows:

> We are a civilian 501(c)(3) [a tax-exempt organization] and that means that we get donations but we still rely on donations. We need money to carry out our mission, so that is a struggle. We are a two-person operation here, and although we have a lot of volunteers and get a lot of pro-bono work as a contribution, it is still a challenge. (24: Non-DoD military family expert)

> Funding. We go through the federal POM [Program Objective Memorandum] process. We have never been funded at more than 22 percent of our validated request, so we are operating at one-fifth capacity. We have been able to make up some of this shortfall—e.g., we have received some GWOT money in the past, and Congress was helpful—very generous—for FY06. But for 2008-2013, we are back to about 24% funding versus validated request. Probably our biggest challenge is to work within those funding limits. (6: DoD military family expert)

> Manning—we don't have enough people given OPTEMPO and the size of the component. (14: DoD military family expert)

The lack of funding often resulted in insufficient paid staff and a reliance on volunteers, which experts viewed as a potential problem in and of itself:

> I've done this job since 2002 and today, our volunteer base, which is the backbone of the organization, is burning out. Since 9/11, our need for volunteers has far exceeded our ability to provide volunteers who are refreshed and willing to give a lot of time. (13: DoD military family expert)

> This system depends too heavily on volunteer support. It is hard to provide consistent services, and programs end up being personality driven and the quality/reliability end up being an issue. (9: DoD military family expert)

It is important to emphasize, however, that while they regretted the need to depend so heavily on unpaid volunteers, the interviewed experts were consistently grateful for the myriad contributions of volunteers. As one expert put it: "There are very committed, passionate volunteers involved in family support. They are truly outstanding" (3: DoD expert). Another individual went perhaps even further by saying, "We need DoD to allow us more money to properly recognize these volunteers" (5: DoD military family expert).

Cross-Leveling and the Resulting Challenges to Family Support

The 2007 report issued by the CNGR found that cross-leveling, or deploying individuals as part of a unit other than that which they had typically trained with,

> has deleterious effects on unit cohesion, training, and readiness and on the ability of the reserve components to provide support to the families of mobilized reservists. One battalion commander testified before the Commission that "cross-leveling is evil." (CNGR, 2007: 19)

In its report, the commission also cites the example of an Army National Guard unit that mobilized with 170 personnel, of which 163 came from a total of "65 units and 49 locations" (CNGR, 2007: 20).

The issue of cross-leveling personnel did occasionally arise during our interviews, either in the context of problems families faced during deployment, their perceptions of military-sponsored support, or in the context of suggestions they made to improve support. Although such comments were unprompted, roughly 8 percent of both spouses and service members raised issues related to deploying separate from one's regular unit. Since we did not expressly ask all spouses and service members to discuss the implications of the service member's deploying with a unit different from the one with which he or she trains, we could not compare the frequency of their responses or look for patterns in the data. We do note, however, that comments of this nature were made by those affiliated with each of the components included in our study, and we mention their remarks because they are consistent with both the results of the CNGR report and with sentiments expressed by the interviewed military family experts. We asked the experts explicitly about Individual Mobilization Augmentees (known also as Individual Augmentees, IMAs, or IAs), but their comments pertain to any service member deployed separate from his or her customary unit. Exemplary comments follow:

> There are also special problems when IAs are mobilized, or when portions of units are mobilized and attached to another unit. Often, nobody is designated to take care of these families. The command becomes responsible for the service member, but they don't always take care of the families. In some cases, a family readiness coordinator or volunteer will call from another state, and say, "I'm your family readiness person, but I'm not sure what I can do for you from this distance." This will continue to be a leadership issue, and given limited resources, the family support issues will be the first thing to fall through. This is why there needs to be a standardization of care for all families. (19: non-DoD military family expert)

> They are the most difficult—they don't belong to a unit. Unit integrity is a really important part of coping for people. It is a group with a common tie—you don't feel alone and this helps people cope. Also, IAs don't get a lot of notice for deployment and they and their families may not know where they will go. This is stressful. And they have no link back to the unit. (12: DoD military family expert)

> Those sent out in one-sies or two-sies face different problems and have different needs. These are the families that are normally isolated from the broader military community and they are the ones that have a harder time accessing resources. They do not have the unit to relate to, and again, this type of "isolation" for the family from a larger community like them can create problems like those I mentioned previously. (8: DoD military family expert)

These experts were concerned primarily about the effect of such individual deployments on the family, which was also the case for spouses and service members, who also tended to refer to the effect of cross-leveling on the family. In the cross-leveling comments, spouses most frequently discussed the lack of connection they felt with the unit their service member had deployed with, and the lack of support they received from that unit's family support. Comments included:

> I don't think you get any real support when the unit that your husband gets cross-leveled in is four or five hundred miles away. I think that cross-leveling is a nightmare, and they should never do it. Because it's very difficult to make families feel like they're included in anything going on in a unit. There's no cohesiveness to it when you're cross-leveled like we were. I didn't even know another wife that was cross-leveled. (383: Army Reserve, E-6's wife)

> My husband was cross-leveled. It would have been nice to have been contacted by a local Family Readiness Group. Just a phone call to say "Hey this is where you can contact us," "Why don't you come to a meeting?" Anything. Anything so I could relate to other wives going through the same situation. I felt entirely isolated and alone. And that put a very bitter taste in my mouth for quite a while after he was deployed. (617: Army Reserve, E-6's wife)

> My husband was deployed on his own, to fill a single slot. So unfortunately, I think we may have fallen through the cracks. We haven't gotten the support from family support. We haven't gotten the phone calls. I don't know if there are newsletters or anything that they send out saying what's going on. I've been able to go online to the Web site and see that they've had some things for children. Unfortunately, I missed them, because I got online too late. (399: Air Force Reserve, O-4's wife)[3]

---

[3]   As mentioned in Chapter One, our interviewees included a small number of service members who had recently been promoted to O-4, and spouses whose service member had recently been promoted. Given challenges in reaching interviewees, we opted to retain such individuals and included them within the O-3 pay grade categories for analysis.

My husband was actually attached to a different unit than what he drilled with, and I never received any information about that group, how I should contact them, what that unit was, information on that unit, or anything. (31: Army National Guard, O-2's wife)

One comment highlighted the inappropriateness of the information the interviewee did receive from the family support of the cross-leveled unit:

[The unit's Family Readiness Group] shouldn't send pamphlets on the local area where the reserve unit is from when in fact the soldier has been cross-leveled and could actually have a family living a thousand miles away. In other words, if I'm crossed-leveled with a unit out of New York and I'm from the state of Washington, don't send me a pamphlet on New York City when I'm living in Washington. The pamphlet on what services are available in New York. I need to see what is available in Washington. (708: Army Reserve, O-4).

Finally, some of the comments pertained to the post-deployment effect on the cross-leveled service members, to include this comment about returning without the support of one's entire unit:

[Our deployed unit was] based on three separate detachments in two separate states and when we came home all units went back to those three separate locations. I was not able to ensure my Marines were ok when we got home, because more than two-thirds of my platoon where no longer in my care. Personal issues or issues upon getting back, trying to fit back in. It's really hard to fly 18 hours out of Iraq and land in the United States and be asked to go home to your house like nothing ever happened was very difficult. [It was difficult] to lose your Marines, because that was your family for a year. They spend 60 days getting us ready to leave but we had about three days to go back to being normal. . . . I was worried about the Marines making good choices when they got home because we're not very patient when we get back. And they're younger and I was very worried about them that way because I could no longer see them. Having your Marines around would have helped me, because even coming home for me was hard and I was much older than they were. (319: Marine Forces Reserve, O-3).

We were not able to quantify the extent to which these families had additional problems or challenges beyond other families. Nonetheless, our research suggests that families of cross-leveled service members may encounter additional difficulties, and may not receive the same level of support as do families whose service members deploy with their usual unit.

Discussion

The results of our interviews indicate that most of the guard and reserve families in our study used some type of resource during their most recent deployment experience. A relatively small number of military-sponsored resources were identified. A wider variety of informal resources were discussed during the interviews, particularly by spouses, who were more inclined than service members to mention turning to an informal resource during deployment. The three most frequently mentioned military-sponsored resources were TRICARE, family support organizations, and Military OneSource, while the three most commonly identified informal resources were extended family, church, and friends and neighbors. Table 7.7 summarizes the factors associated with the mention of each of these resources and indicates whether each relationship was found in the service member portion of the sample, the spouse portion, both, or neither.

Several indicators of maturity or being established were related to a tendency to use either military or nonmilitary resources. Older service members were more inclined to mention their family's use of TRICARE, and older spouses were more likely than younger spouses to turn to a religious organization or their friends and neighbors for assistance during deployment. Newlyweds were generally less inclined to use support resources, with newly married service members citing TRICARE less often and newly married spouses less frequently dis-

Table 7.7
Summary of Factors Related to Resource Usage

	Military			Informal		
	TRICARE	Family Support Organizations	Military One-Source	Family	Religious Organizations	Friends, Neighbors
Individual and situational characteristics						
Age	SM				SP	SP
Gender	SM	SM				
Marriage length	SM		SP		SP	SP
Parental status	SP		SP		SM	
College degree		SM		SP	SM	SP, SM
Service member pay grade		SP, SM	SP, SM	SP	SP	SP, SM
Service member reserve component	SP, SM	SP, SM	SP, SM	SP	SM	SP
Service member prior active duty		SP	SP	SM		
Spouse prior military (spouses only)				SP		
Financial situation			SP, SM	SP		SP
Distance from drill unit		SP			SP, SM	
Distance from nearest military installation				SP	SP, SM	
Amount of notice	SM	SP		SP		
Perception of notice adequacy			SP	SM		SM
Deployment length	SP, SM	SP, SM	SP, SM		SP, SM	
Family readiness		SM	SP			SP

SOURCE: 2006 RAND Guard and Reserve Family Interviews.

NOTES: All relationships listed are statistically significant at p<0.10. SM = Finding present in the service member portion of the sample (N=296). SP = Finding present in the spouse portion of the sample (N=357).

cussing Military OneSource, church, or friends and neighbors than their counterparts. On the other hand, having children and possessing a college degree tended to be related to more frequently identifying resources; those with children tended to mention TRICARE and Military OneSource, while those with a college degree were more likely to use family support organizations and all three types of informal resources. One exception was that service members *without* children were more inclined to discuss their spouse's turning to a religious organization for support during their employment. Family finances suggest explanations for resource usage as well. Families with a comfortable financial situation were more likely to use extended family than were those with uncomfortable finances, while those with uncomfortable finances were more likely to mention the use of Military OneSource and friends and neighbors.

Service member gender was another factor related to the use of military resources. Specifically, female service members were less likely to state that their family used TRICARE or family support organizations. Perhaps their family obtained health insurance via their husband's employer instead, and, as suggested by the military family experts we interviewed, husbands may be less inclined to access the family readiness programs offered to them. Service member pay grade and reserve component also helped to explain families' use of military and informal resources. Junior enlisted personnel and spouses were more likely to cite their family's use of family support organizations and Military OneSource. In addition, junior officer spouses were more inclined to mention turning to extended family and religious organizations than were other spouses, and both junior officers and spouses of junior officers mentioned friends and neighbors more often than did their counterparts in other pay grades. With respect to reserve component, Army National Guard families were more likely to turn to TRICARE, family support organizations, Military One-Source, and religious organizations for support than were families from at least one, and frequently more, of the other reserve components. In addition, Army Reserve spouses tended to mention TRICARE and family support organizations more often than did Air Force Reserve and Marine Forces Reserve spouses. Marine Forces Reserve spouses were more inclined to describe how their extended family helped them during deployment, while Marine reservists were less inclined to cite Military OneSource as a resource for their family. Lastly, Air Force Reserve spouses rarely mentioned Military OneSource, but more frequently identified friends and neighbors as a resource than did Army Reserve spouses.

Usage of military and informal resources was also related to multiple measures of experience. Prior military experience tended to be associated with less frequent mention of various support resources; families of service members with prior active duty experience were less likely to use family support organizations, Military OneSource, and extended family, and spouses with prior military experience were less inclined to discuss extended family than were those without military experience.

Consistent with comments made by military experts during their interviews, families' distance from either the drill unit or the nearest military installation was related to their use of resources as well. Families residing farther away from the drill unit were less inclined to use family support organizations and more inclined to rely on religious organizations for support than were those living closer to the drill unit. In addition, those farther away from a military installation were more inclined to turn to church or other religious organizations than were those living closer to a military installation. Spouses living closer to the nearest military installation, in contrast, were more inclined to discuss support provided by extended family.

Lastly, characteristics of the most recent deployment appeared to influence family's use of support resources. Less actual notice was associated with more frequent mention of TRICARE, while those receiving more notice were more inclined to mention family support organizations and extended family. Perceptions of insufficient notice were related to more frequent reference to Military OneSource and to friends and neighbors, as well as to less frequent reference to extended family, all in comparison with those who deemed their notice adequate. In addition, those who perceived their family as very ready for deployment were more inclined to cite their use of family support organizations and their reliance on friends and neighbors than those whose families were not ready at all, and they were also less inclined to mention Military OneSource. It is unclear however, whether usage of family support organizations and friends and neighbors affected their level of readiness, or if their high level of readiness influenced their decision to turn to these forms of support. Finally, both service members and spouses who experienced longer deployments tended to report the use of all three types of military resources as well as a reliance on their religious organization.

None of the military-sponsored resources was cited by a majority of spouses or service members, and, with the exception of extended family, the same finding also pertained to informal resources. Our interviews with spouses and service members, as well as those with military family experts, suggested some possible reasons for this result. Specifically, spouses who did not use military-sponsored support programs most frequently said that this was because they had no need for such services. Other spouses described accessibility-related challenges stemming from distance, time, or child-related constraints, and a small number of spouses reported not using military resources because they were of insufficient quality. Comments made by spouses and service members suggest that informal resources may not have been accessed by larger proportions because of a lack of awareness. Military family experts' remarks corroborated this premise, and the interviewed experts discussed how families' geographic dispersal and difficulties actually contacting them could contribute to this lack of awareness and subsequent use. The experts also identified a lack of funding as another obstacle to successfully reaching and supporting reserve component families.

CHAPTER EIGHT

How Do Guard and Reserve Families' Retention Plans Differ?

In earlier sections of this monograph, we discussed family perceptions of the notice they received prior to activation, family readiness, the problems and positives that service members and spouses associated with deployment, and family coping. In this chapter, we consider the implications that those findings potentially have for the retention of guard and reserve personnel. Three questions related to retention intention were included in our interviews with service members: intentions to stay until retirement eligibility, the impact of the most recent activation on the service member's career plans, and his or her spouse's opinion toward the service member's military career. Comparable versions of the intent to stay and spouse opinion questions were posed to spouses themselves as well. Service members' and spouses' responses to these items, along with patterns significantly related to both favorable and unfavorable retention implications, are summarized herein to demonstrate that family characteristics and experience may come to bear on one indicator of military effectiveness, retention.

Intentions to Stay Until Retirement Eligibility

Using a question adapted from the 2006 Survey of Reserve Component Spouses, we asked both spouses and service members to characterize the service member's career plans by selecting one of five statements pertaining to career length. The options ranged from leaving before the present military obligation was completed through staying until the mandatory retirement age was reached. For the purposes of our analysis, these statements were consolidated into two measures: leave before retirement eligibility and stay until retirement eligibility. About 4 percent of service members and 8 percent of spouses did not provide a response to this question; these individuals often indicated they did not have a high enough degree of certainty to respond. Responses of those who did estimate the service member's career length are depicted in Figure 8.1. Although the service members and spouses were not married to one another and typically came from different units, their responses are very similar. Just over half of both—54 percent of service members and 52 percent of spouses—indicated plans for the service member to stay in the guard or reserve until retirement eligibility. On the other hand, 42 percent of service members and 41 percent of

Figure 8.1
Summary of Intentions to Stay Until Retirement Eligibility

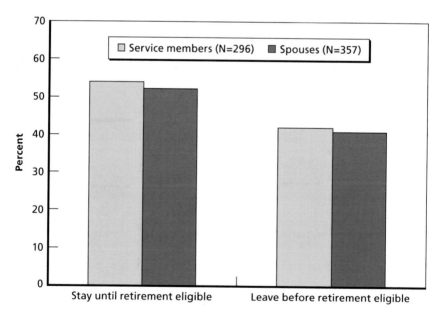

SOURCE: 2006 RAND Guard and Reserve Family Interviews.
RAND *MG645-8.1*

spouses indicated plans to leave before the service member qualified for retirement.

Spouses and service members had some similarities in how they responded to this question, but varied in a number of ways as well. Statistically significant patterns for both groups of interviewees are provided in Table 8.1. This table is similar to ones featured in earlier chapters, but also includes significant patterns related to the problems discussed in Chapter Four and the positives described in Chapter Five. For both, marriage and age were positively related to an increased likelihood to stay until retirement eligibility; older spouses and service members and those with longer marriages were more inclined to indicate plans to stay until retirement eligibility than were younger individuals and those in newer marriages. For example, 33 percent of service members and 37 percent of spouses who were newlyweds reported plans to stay until retirement eligibility, compared with 61 percent of both service members and spouses married for three years or more. Junior enlisted personnel and those married to junior enlisted personnel were also less likely to plan to stay until retirement eligibility, or, conversely, more likely to plan to leave before retirement eligibility than their counterparts in mid-grade enlisted and junior officer pay grades. In addition, Air Force reservists were more likely than Marine reservists to plan to stay until retirement eligibility; 63 percent of Air Force reservists planned to stay to this point, compared with 47 percent of Marine reservists. A similar yet broader finding was present within the spouse sample: Air Force Reserve spouses were more likely than spouses from the Army National Guard, Army Reserve, and Marine Forces Reserve to note plans for their service member to stay until retirement eligibility.

Spouses and service members also resembled one another in how perceptions of notice adequacy and family readiness came to bear on plans to stay until retirement. For both, those who felt the notice received was adequate and those who described their family as ready or very ready were more inclined to stay than were those who deemed the notice insufficient and those who felt that their family was not prepared at all. To illustrate, 61 percent of service members and 65 percent of spouses who described the notice as adequate planned to stay until

Table 8.1
Characteristics Associated with Service Member Career Plans

	Service Members (%)		Spouses (%)	
	Leave Before Retirement Eligibility	Stay Until Retirement Eligibility	Leave Before Retirement Eligibility	Stay Until Retirement Eligibility
Age				
25 or less (N=36 service members; N=49 spouses)	81	19	76	25
26 or more (N=248 service members; N=281 spouses)	39	61	39	61
Gender				
Male (N=260)	42	58		
Female (N=25)	60	40		
Marriage length				
2 years or less (N=48 service members; N=67 spouses)	67	33	63	37
3 years or more (N=217 service members; N=263 spouses)	39	61	40	61
Parental status				
Has children (N=223)	40	60		
No children (N=62)	57	44		
College degree				
Yes (N=180)			39	61
No (N=150)			50	50
Service member pay grade				
E-1 to E-4 (N=63 service members; N=83 spouses)	62	38	68	33
E-5 to E-6 (N=141 service members; N=163 spouses)	38	62	38	62
O-1 to O-3 (N=81 service members; N=84 spouses)	40	61	33	67
Service member reserve component				
Army National Guard (N=100 service members; N=95 spouses)	42	58	45	55
Army Reserve (N=70 service members; N=82 spouses)	44	56	55	45

Table 8.1—Continued

	Service Members (%)		Spouses (%)	
	Leave Before Retirement Eligibility	Stay Until Retirement Eligibility	Leave Before Retirement Eligibility	Stay Until Retirement Eligibility
Air Force Reserve (N=59 service members; N=79 spouses)	37	63	20	80
Marine Forces Reserve (N=56 service members; N=74 spouses)	54	47	57	43
Service member prior active duty				
Yes (N=192)			35	65
No (N=138)			57	44
Financial situation				
Comfortable (N=187)	40	60		
Occasional difficulty (N=64)	45	55		
Uncomfortable (N=34)	65	35		
Perception of notice adequacy				
Adequate (N=130 service members; N=148 spouses)	39	61	35	65
Insufficient (N=50 service members; N=59 spouses)	58	42	56	44
Family Readiness				
Ready or very ready (N=185 service members; N=198 spouses)	38	62	37	63
Somewhat ready (N=43 service members; N=66 spouses)	58	42	53	47
Not at all ready (N=47 service members; N=51 spouses)	60	40	55	45
Problems				
Emotional or mental				
Cited problem (N=132)			52	48
Did not cite problem (N=198)			39	61

Table 8.1—Continued

	Service Members (%)		Spouses (%)	
	Leave Before Retirement Eligibility	Stay Until Retirement Eligibility	Leave Before Retirement Eligibility	Stay Until Retirement Eligibility
Education				
Cited problem (N=39)	56	44		
Did not cite problem (N=246)	42	58		
Health care				
Cited problem (N=38)			61	40
Did not cite problem (N=292)			42	58
No problems				
Cited "no problems" (N=49)			29	71
Did not cite "no problems" (N=281)			47	53
Positives				
Financial gain				
Cited positive (N=76)	33	67		
Did not cite positive (N=209)	48	52		
Family closeness				
Cited positive (N=59)	29	71		
Did not cite positive (N=226)	48	52		
No positives				
Cited "no positives" (N=54)	70	30		
Did not cite "no positives" (N=231)	38	62		

SOURCE: 2006 RAND Guard and Reserve Family Interviews.

NOTES: Ns are provided for either service member or spouse, as denoted in the table. When data from both groups are shown, Ns are specified as service member or spouse. All percentages shown are statistically different from one another at p<0.10. Shading indicates a subset of population that is not significantly different from other subsets. For pay grade comparisons in both the service member and the spouse groups, the E-1 to E-4 category is significantly different from the E-5 to E-6 and O-1 to O-3 categories. The other pay grade comparison is not significantly different. For reserve component comparisons in the spouse group, the Air Force Reserve is significantly different from the three other components. Other reserve component comparisons are not significantly different.

retirement eligibility, as did 62 percent of the service members and 63 percent of the spouses who characterized their family as ready or very ready.

Turning our attention toward differences between the spouses and service members, we found many instances in which there was a notable finding present for only one of the two groups. It is not clear, however, whether these differences stem from the unique experiences of the spouses and service members in our sample or are indicative of a divergence between spouse and service member perceptions of similar situations. Among the service members, male service members, those with children, and those with more comfortable family finances were more inclined to stay. Patterns based on gender, parental status, or family finances were not present in the spouse data, nor was there a relationship between any of the positive aspects of deployment spouses cited and plans to stay until retirement. In addition, service members who cited financial gain or family closeness as positives were more inclined to stay until retirement than those who did not mention those positive aspects. Further, not being able to recall any positive aspects was related to a higher tendency for service members to leave before retirement eligibility, and this was the case for service members who mentioned education-related problems as well.

Among the spouses, there was a negative relationship between health care–related problems and plans to stay until retirement eligibility that was not evident for the service members we interviewed. Sixty-one percent of the spouses that cited this type of problem indicated service member plans to leave before retirement eligibility, while only 42 percent of those who did not refer to this problem expressed similar career plans. References to emotional or mental problems also had similarly negative implications for planned career length. In addition, *not* identifying any problems was positively related to plans to stay until eligible for retirement. Seventy-one percent of spouses who indicated their family did not experience any problems stemming from deployment also expressed service member plans to stay until retirement eligibility, compared with 53 percent of spouses who mentioned a problem of any type during their interview. Lastly, college-educated spouses and those married to service members with prior active duty experience were more likely to report plans to stay until retirement than were their less educated and less experienced counterparts, respectively. This latter finding may reflect more experience with deployments and military life or more simply the fact that the service member has more years accrued toward retirement eligibility.

The Impact of the Most Recent Activation on Service Member Career Plans

During their interviews, service members were asked, "What impact has your recent activation had on your [National Guard/Reserve] career intentions?" Army guardsmen were asked about their National Guard career intentions, while members of the other three included reserve components were asked about Reserve career intentions. A five-point scale was provided, which ranged from "greatly increased my desire to stay" to "greatly increased my desire to leave," and, for ease of presentation, was collapsed into a three-point scale for our analysis. A breakdown of service members' responses is provided in Figure 8.2. Thirty-eight percent of service members said their most recent activation had no influence on their career plans, while comparable proportions of service members indicated it either increased their desire to stay or increased their desire to leave (30 percent and 32 percent, respectively).

Our analysis indicated that a number of characteristics, problems, positives, family readiness, and family coping served as the basis for statistically significant patterns related to this question. A list of these patterns and the response frequencies associated with each is provided in Table 8.2. Female service members, younger service members, and those with less comfortable family finances were all more inclined to offer negative views of their most recent activation. Female service members, for instance, were more likely to have negative views of their most recent activation; 54 percent of them said it increased their desire to leave, compared with 30 percent of male service members. Service members' pay grades also were related to their opinion; junior enlisted personnel were more inclined to say that the most recent

Figure 8.2
Summary of the Impact of the Most Recent Activation on Service Member Career Plans

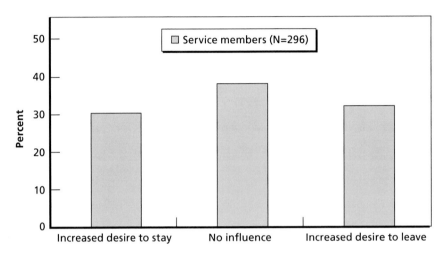

SOURCE: 2006 RAND Guard and Reserve Family Interviews.
RAND *MG645-8.2*

activation increased their desire to leave than were junior officers (44 percent versus 25 percent, respectively). There were differences present among the reserve components: Army National Guard personnel and Army reservists were more likely than Air Force reservists to state that their most recent activation increased their desire to leave. Thirty-eight percent of Army guardsmen and 39 percent of Army reservists felt this way, compared with 19 percent of Air Force reservists. This may be related to differences in deployment length or the amount of notice; each of these factors was also associated with service members' view of their most recent activation. Specifically, the longer the most recent deployment, the more likely service members were to note their most recent activation increased their desire to leave. In the case of notice, the longer the notice, the more likely service members were to indicate that their most recent activation increased their desire to *stay* (or conversely, were less likely to say it increased their desire to leave). Since, as shown in Chapter Two, average deployments for Army National Guard and Army Reserve service members in our sample were greater than

Table 8.2
Characteristics Associated with the Impact of the Most Recent Activation on Service Member Career Plans

	Service Members (%)		
	Increased or Greatly Increased Desire to Stay	No Impact	Increased or Greatly Increased Desire to Leave
Age			
25 or less (N=37)	39	14	49
26 or more (N=256)	29	41	30
Gender			
Male (N=268)	31	38	30
Female (N=26)	15	31	54
Service member pay grade			
E-1 to E-4 (N=69)	29	28	44
E-5 to E-6 (N=144)	30	39	31
O-1 to O-3 (N=81)	31	44	25
Service member reserve component			
Army National Guard (N=104)	26	37	38
Army Reserve (N=74)	26	35	39
Air Force Reserve (N=58)	31	50	19
Marine Forces Reserve (N=58)	41	31	28
Financial situation			
Comfortable (N=194)	32	41	27
Occasional difficulty (N=65)	32	35	32
Uncomfortable (N=35)	14	26	60
Amount of notice			
One month or less (N=155)	23	37	40
More than one month (N=139)	38	38	24
Perception of notice adequacy			
Adequate (N=132)	43	36	21
Insufficient (N=54)	15	20	65

Table 8.2—Continued

	Service Members (%)		
	Increased or Greatly Increased Desire to Stay	No Impact	Increased or Greatly Increased Desire to Leave
Deployment length			
Less than one year (N=122)	35	40	25
One year or more (N=172)	26	36	38
Family Readiness			
Ready or very ready (N=191)	36	41	24
Somewhat ready (N=42)	21	38	41
Not at all ready (N=51)	16	24	61
Problems			
Employment			
Cited problem (N=61)	25	30	46
Did not cite problem (N=233)	31	40	29
Education			
Cited problem (N=41)	10	34	56
Did not cite problem (N=253)	33	28	29
Marital			
Cited problem (N=36)	11	39	50
Did not cite problem (N=258)	33	38	29
Health care			
Cited problem (N=35)	17	29	54
Did not cite problem (N=259)	32	39	29
No problems			
Cited "no problems" (N=87)	40	39	20
Did not cite "no problems" (N=207)	26	37	37

Table 8.2—Continued

	Service Members (%)		
	Increased or Greatly Increased Desire to Stay	No Impact	Increased or Greatly Increased Desire to Leave
Positives			
Family closeness			
Cited positive (N=60)	47	40	13
Did not cite positive (N=234)	26	37	37
No positives			
Cited "no positives" (N=60)	15	25	60
Did not cite "no positives" (N=234)	34	41	25
Family coping			
Coped well (N=185)	35	38	27
Coped moderately (N=47)	23	40	36
Coped poorly (N=25)	16	32	52

SOURCE: 2006 RAND Guard and Reserve Family Interviews.

NOTES: Ns are provided for either service member or spouse, as denoted in the table. When data from both groups are shown, Ns are specified as service member or spouse. All percentages shown are statistically different from one another at p<0.10. Shading indicates a subset of population that is not significantly different from other subsets. For reserve component comparisons, the Air Force Reserve is significantly different from the Army National Guard and the Army Reserve. Other reserve component comparisons are not significantly different.

one year, and over two-thirds of Army reservists received one month's notice or less, these findings may help to explain the reserve component patterns and may guide future efforts to better understand the interrelationships of these factors.[1]

Not only was actual notice related to this measure of retention intention, as discussed above, but additionally, perceptions that the

---

[1]   As noted in Chapter One and Appendix B, given both the exploratory nature of our study and resources available, our statistical analysis focused on exploring bivariate relationships (those between two measures). We concur with one of our reviewers that efforts in which these relationships are considered simultaneously (i.e., multivariate analyses) would be of value to those seeking to determine the relative magnitude and significance of these factors.

notice was adequate were related to favorable views of the most recent activation, and the reverse was true for notice described as insufficient. Specifically, 43 percent of service members who felt the notice adequate noted that their most recent activation increased their desire to stay, compared with 15 percent of those who regarded the amount of notice as insufficient. Conversely, 65 percent of service members who felt the notice was insufficient said the activation increased their desire to leave, while only 21 percent of service members who perceived their notice as adequate felt that way. A similar relationship was apparent for family readiness as well: Service members who believed that their family was ready or very ready tended to have a favorable opinion of their most recent activation, and those who said that their family was completely unprepared tended to view it in a more unfavorable light.

Moreover, aspects of the deployment itself, namely the problems and positives cited by personnel, and how well families coped, appear to influence service member perceptions of their most recent activation. Service members who did not mention any problems and those who cited family closeness as a positive were both more inclined to report their most recent activation increased their desire to stay. Forty percent of those who did not identify problems said their activation increased their desire to stay, compared with 26 percent of those who did discuss at least one problem. In addition, many of the problems cited by service members were related to an increased tendency to view the most recent activation unfavorably; service members with problems related to employment, education, marriage, and health care all were more likely to state that the most recent activation increased their desire to leave than were their counterparts without that specific problem. For example, 50 percent of service members who had marital problems felt that their most recent activation increased their desire to leave, while 29 percent of service members who did not cite marital problems expressed a similar sentiment. Failing to mention any positive aspects of deployment was also related to unfavorable views of the most recent activation. Finally, there was a significant relationship between family coping (the focus of Chapter Six) and service members' views of their most recent activation. Specifically, 35 percent of service members who felt that their family coped well or very well during their deployment also indicated their activation increased their desire to stay, compared with only 16 percent of those who reported that their family coped poorly.

Spouse Opinion Regarding Service Member Career Plans

The final measure of retention intention was spouse opinion regarding whether the service member should stay or leave. This question was adapted from one posed to active duty personnel in the July 2005 Status of Forces Survey for Active Duty Members. Specifically, spouses in our study were asked whether they thought their service member should stay in or leave the National Guard/Reserve, and service members were asked a similar question with respect to the views of their spouse or significant other. Response options ranged on a five-point scale from strongly favors staying to strongly favors leaving, and as with the other retention-related measures, the responses were converted to a three-point scale during our analysis. In this case, almost all spouses gave an answer (98 percent), but service members had a slightly harder time assessing their spouse's opinion: 9 percent of them did not provide an answer. The views of those who did respond to the question are illustrated in Figure 8.3. In contrast to Figure 8.1, depicting retirement plans, spouses and service members varied in their views overall, and these differences were statistically significant. Fifty-eight percent of spouses favored their service member's staying in the Guard or Reserve, compared with just 35 percent of service members who believed that their spouse favored their staying. In addition, 40 percent of service members indicated that their spouse favored their leaving, while only 25 percent of the spouses in our study stated they felt that way. Given that we did not interview the spouses of the service members in our sample, however, it is unclear, whether the individuals married to the service members in our study actually had less favorable views or if that was the service members' perception of their views.

Figure 8.3
Summary of Spouse Opinion Regarding Service Member Career Plans

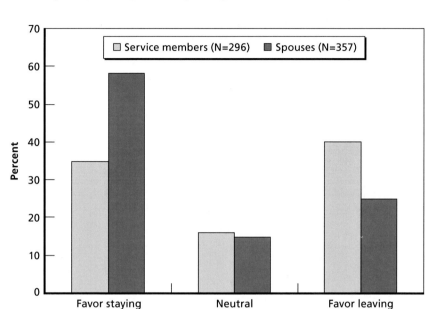

SOURCE: 2006 RAND Guard and Reserve Family Interviews.
RAND *MG645-8.3*

We considered spouse opinion a possible indirect measure of service member career intentions, and statistical analysis[2] revealed a significant relationship between service members' career plans and spouses' opinions. Specifically, a favorable spouse opinion (i.e., favoring staying) was associated with plans to stay until retirement eligibility, and an unfavorable spouse opinion (i.e., favoring leaving) was associated with plans to leave before reaching that career milestone. This was true for both the spouse and service member portions of the sample. For example, when a spouse indicated that she favored her service member staying, she also tended to indicate that he planned to stay until retirement eligibility. Likewise, when a service member indicated that his

---

[2]   We used regression analysis to examine the relationship between these two measures of retention intentions. Results were significant at $p<0.10$ for both the spouse and service member groups.

spouse favored him leaving, he tended to state that he planned to leave the Guard or Reserve before retirement eligibility.

As with plans to stay until retirement, some patterns of responses were common among spouses and service members. These relationships, as well as those unique to the service member or spouse portions of the interview sample, are summarized in Table 8.3. Air Force Reserve personnel and spouses were again somewhat different from other components in their responses. Among service members, Air Force reservists were more inclined to say that their spouse favored their staying than were Army or Marine reservists. Spouses of Air Force reservists were more likely than spouses from the Army National Guard, Army Reserve, or Marine Forces Reserve to say that they favored the reservist staying. Marriage length was another factor related to spouse opinion; those in longer marriages were more inclined to express a spouse opinion that favored the service member's remaining in the Guard or Reserve. Spouses and service members were also similar in how their perceptions of family readiness were related to spouse opinion. Both spouses and service members whose families were ready or very ready tended to note that they and their spouses, respectively, favored the service member staying in the Guard or Reserve.

In addition, emotional/mental and marital problems were both associated with spouse opinions: Service members and spouses who mentioned these problems were more inclined to provide an unfavorable spouse opinion (i.e., a preference for the service member leaving the Guard or Reserve). This was also true for spouses and service members who could not recall any positive aspects of deployment. To illustrate, 59 percent of service members and 48 percent of spouses who could not identify any positives expressed unfavorable spouse opinions, compared with 40 percent of service members and 22 percent of spouses who did identify at least one positive aspect during their interview. Conversely, failing to mention any problems was related to more favorable spouse opinions (i.e., a preference for the service member staying in the Guard or Reserve). Fifty-one percent of the service members who did not cite any problems, for example, said that their spouse favored their staying, while just 33 percent of service members who did identify problems expressed a similar viewpoint. Recognizing the financial benefits of

Table 8.3
Characteristics Associated with Spouse Opinion Regarding Service Member Career Plans

	Service Members (%)			Spouses (%)		
	Favors Staying	Neutral	Favors Leaving	Favors Staying	Neutral	Favors Leaving
Overall percentages (N=269 SM; N=352 SP)	39	18	44	59	15	26
Age						
25 or less (N=33)	18	21	61			
Age 26 or more (N=235)	41	17	41			
Gender						
Male (N=247)	41	18	42			
Female (N=22)	18	18	64			
Marriage length						
2 years or less (N=50 SM; N=69 SP)	24	22	54	42	17	41
3 years or more (N=219 SM; N=283 SP)	42	17	41	63	15	22
Service member pay grade						
E-1 to E-4 (N=61)	25	21	54			
E-5 to E-6 (N=130)	42	17	41			
O-1 to O-3 (N=78)	44	17	40			
Service member reserve component						
Army National Guard (N=99 SM; N=101 SP)	41	19	39	54	17	30
Army Reserve (N=70 SM; N=88 SP)	30	17	53	55	15	31
Air Force Reserve (N=51 SM; N=83 SP)	53	18	29	76	12	12
Marine Forces Reserve (N=49 SM; N=80 SP)	31	16	53	54	16	30

Table 8.3—Continued

	Service Members (%)			Spouses (%)		
	Favors Staying	Neutral	Favors Leaving	Favors Staying	Neutral	Favors Leaving
Financial situation						
Comfortable (N=182 SM)	45	18	37			
Occasional difficulty (N=57 SM)	33	19	47			
Uncomfort-able (N=30 SM)	13	13	73			
Distance from drill unit						
Less than 25 miles (N=75)	48	12	40			
25 or more miles (N=194)	35	20	45			
Distance from nearest military installation						
Less than 100 miles (N=224)	42	18	41			
100 or more miles (N=45)	24	18	58			
Perception of notice adequacy						
Adequate (N=122)	49	12	39			
Insufficient (N=47)	17	17	66			
Repeat OCONUS deployments since 9/11						
No (N=221)				63	13	24
Yes (N=131)				52	19	29
Family Readiness						
Ready or very ready (N=177 SM; N=212 SP)	47	16	37	65	15	20
Somewhat ready (N=40 SM; N=68 SP)	33	23	45	53	7	40
Not at all ready (N=44 SM; N=55 SP)	9	21	71	51	24	26

Table 8.3—Continued

	Service Members (%)			Spouses (%)		
	Favors Staying	Neutral	Favors Leaving	Favors Staying	Neutral	Favors Leaving
Problems						
Emotional or mental						
Cited problem (N=71 SM; N=139 SP)	35	11	54	50	14	35
Did not cite problem (N=198 SM; N=213 SP)	40	20	40	65	16	20
Marital						
Cited problem (N=26 SM; N=38 SP)	12	19	69	42	16	42
Did not cite problem (N=243 SM; N=314 SP)	42	18	41	61	15	24
No problems						
Cited "no problems" (N=83 SM; N=49 SP)	51	22	28	65	25	10
Did not cite "no problems" (N=186 SM; N=303 SP)	33	16	51	58	14	28
Positives						
Financial gain						
Cited positive (N=71 SM; N=72 SP)	54	10	37	71	10	19
Did not cite positive (N=198 SM; N=280 SP)	33	21	46	56	16	28
Family closeness						
Cited positive (N=56)	63	14	23			

Table 8.3—Continued

	Service Members (%)			Spouses (%)		
	Favors Staying	Neutral	Favors Leaving	Favors Staying	Neutral	Favors Leaving
Did not cite positive (N=213)	32	19	49			
Pride, patriotism						
Cited positive (N=84)				71	10	19
Did not cite positive (N=268)				55	17	28
No positives						
Cited "no positives" (N=54 SM; N=48 SP)	13	28	59	35	17	48
Did not cite "no positives" (N=215 SM; N=304 SP)	45	15	40	63	15	22
Family coping						
Coped well (N=217)				65	16	19
Coped moderately (N=70)				51	13	36
Coped poorly (N=25)				36	12	52

SOURCE: 2006 RAND Guard and Reserve Family Interviews.

NOTES: SM = service members; SP = spouses. Ns are provided for either service member or spouse, as denoted in the table. When data from both groups are shown, Ns are specified as service member or spouse. All percentages shown are statistically different from one another at $p<0.10$. Shading indicates a subset of population that is not significantly different from other subsets. For pay grade comparisons in the service member group, the E-1 to E-4 category is significantly different from the E-5 to E-6 and O-1 to O-3 categories. The other pay grade comparison is not significantly different. For reserve component comparisons in the spouse group, the Air Force Reserve is significantly different from the Army Reserve and Marine Forces Reserve, and they are not statistically different from one another. For reserve component comparisons in the spouse group, the Air Force Reserve is significantly different from the three other components. Other reserve component comparisons are not significantly different.

deployment was also linked with favorable spouse opinions for both the spouses and service members in our study.

There were also some unique findings that were apparent only among the service members or the spouses in our study, but no contradictory findings were evident. For service members, age had a positive relationship with spouse opinion; older service members were more likely to say their spouse favored their staying. A similar finding was noted for comfortable family finances. In addition, once again the responses of junior enlisted personnel had negative implications for retention compared with those of their mid-grade enlisted and junior officer counterparts; spouses of junior enlisted personnel in our sample were more inclined to favor their service member leaving. Fifty-four percent of junior enlisted service members stated that their spouse favored their leaving, compared with 41 percent of mid-grade enlisted service members and 40 percent of junior officers.

Moreover, distance from both the drill unit and the nearest military installation was associated with spouse opinion. Service members closer to their drill unit or a military base were more inclined to say their spouse favored their staying than were service members farther away from either location. We also found that husbands of female service members were more likely to favor their spouse leaving (according to the service member) than were wives of male service members. Put another way, only 18 percent of female service members stated that their spouse favored their staying, compared with 41 percent of male service members who expressed a similar sentiment. Although the number of husbands in the service member portion of our sample was small, this finding may have broader implications for the growing number of female personnel in the Guard and Reserve. Those who felt their notice was insufficient also tended to mention unfavorable spouse opinions. Finally, service members who identified family closeness as a positive aspect of deployment tended to believe that their spouse favored their staying.

The spouse portion of the interview sample had notably fewer unique patterns that were statistically significant. Spouses whose service member had only deployed once since 9/11 were more inclined to favor their service member remaining in the Guard or Reserve. With respect to the positive aspects of deployment, spouses who referred to feelings of patriotism, pride, or civic responsibility as a positive also tended to favor their service member staying: 71 percent of spouses who mentioned this positive aspect favored their service member staying, compared with 55 percent of those who did not cite it. Lastly, there was a strong relationship between family coping and spouse opinion: The better the family coped with the deployment, the more favorable the spouse opinion. On the other hand, the worse the family coped, the more unfavorable the spouse opinion. Sixty-five percent of spouses whose felt that their family coped well or very well favored their service member staying, versus 36 percent of those whose families coped poorly. Further, 52 percent of spouses who indicated their family coped poorly favored their service member leaving, compared with just 19 percent of those who believed their family coped well.

Discussion

In this chapter we provided evidence supporting the premise that family attributes and family perceptions of their deployment experience may come to bear on one indicator of military effectiveness. Specifically, we examined how individual and family characteristics, family readiness, the problems and positives stemming from deployment, and family coping relate to multiple measures of retention intentions. Our analysis was based on cross-sectional data, however, so we could not determine whether these various factors have direct effects on retention intentions; it is possible, for example, that an underlying personal attribute not assessed in this study could exert an influence on both family readiness and military career plans. Table 8.4 lists the various factors associated with at least one of the three measures of retention intentions: plans to stay until retirement eligibility, impact of the most recent activation, and spouse opinion regarding the service member's career plans. Whether the finding pertains to the service member portion of the interview sample, the spouse portion, or both is also shown. For example, patterns related to age and intention to stay until retirement eligibility were present among both the spouses and service mem-

Table 8.4
Summary of Factors Related to Retention Intentions

	Intentions to Stay Until Retirement Eligibility	Impact of the Most Recent Activation on Service Member Career Plans (Service Members Only)	Spouse Opinion Regarding Service Member Career Plans
Individual and Situational Characteristics			
Age	SM, SP	SM	SM
Gender	SM	SM	SM
Marriage length	SM, SP		SM, SP
Parental status	SM		
College degree	SP		
Service member pay grade	SM, SP	SM	SM
Service member reserve component	SM, SP	SM	SM, SP
Service member prior active duty	SP		
Financial situation	SM	SM	SM
Distance from drill unit			SM
Distance from nearest military installation			SM
Amount of notice		SM	
Perception of notice adequacy	SM, SP	SM	SM
Deployment length		SM	
Repeat OCONUS deployments			SP
Family readiness	SM, SP	SM	SM, SP
Problems			
Emotional or mental	SP		SM, SP
Employment		SM	
Education	SM	SM	
Marital		SM	SM, SP
Health care	SP	SM	

Table 8.4—Continued

	Intentions to Stay Until Retirement Eligibility	Impact of the Most Recent Activation on Service Member Career Plans (Service Members Only)	Spouse Opinion Regarding Service Member Career Plans
No problems	SP	SM	SM, SP
Positives			
Financial gain	SM		SM, SP
Family closeness	SM	SM	SM
Pride, patriotism			SP
No positives	SM	SM	SM, SP
Family Coping		SM	SP

SOURCE: 2006 RAND Guard and Reserve Family Interviews.

NOTES: All relationships listed are statistically significant at p<0.10. SM = Finding present in the service member portion of the sample (N=296). SP = Finding present in the spouse portion of the sample (N=357).

bers in our interview sample, but for spouse opinion regarding service member career plans, a relationship between age and this measure of retention intentions was only apparent among the service members we interviewed.

Table 8.4 includes factors with favorable and unfavorable implications for service member career plans. Those more likely to have favorable retention plans, suggested by either plans to stay until retirement eligibility, or spouse opinion favoring staying, include older service members and spouses, those in longer marriages, service members with children, spouses with a college degree, and service members with comfortable family finances. Military-related characteristics also played a role: Mid-grade enlisted personnel and spouses of mid-grade enlisted personnel, junior officers and spouses of junior officers, Air Force Reserve personnel and spouses, spouses married to service members with prior active duty experience, and service members living closer to their drill unit and/or the nearest military installation were more likely to have favorable retention intentions. Factors associated with deployment perceptions and experience were also significantly related to retention intentions: Service members receiving more activation notice, spouses and service members who perceived the notice to be adequate, service members who experienced shorter deployments, and spouses whose service member had only been deployed OCONUS once since 9/11 all were more likely to express a preference for the service member to remain in the Guard or Reserve. In addition, those who did not report *any* problems or that cited positive aspects of deployment, such as financial gain; family closeness; or patriotism, pride, or civic responsibility, tended to have more positive views toward retention as well. Finally, those whose families were ready or very ready for the deployment and those whose families coped well were more inclined to stay, as indicated by their responses to the retention intention questions.

Those more likely to have unfavorable retention plans, denoted by either plans to leave before retirement eligibility, an increased desire to leave after the most recent activation, or spouse opinion favoring leaving, include many with the opposite characteristics, such as younger service members and spouses. Additionally, junior enlisted personnel and spouses were more likely to have unfavorable views toward retention, as were female service members and male spouses married to the female service members in our study. Many of the problems mentioned in Chapter Four were negatively related to retention intentions, as was failing to cite any positive aspects of deployment. Lastly, those who characterized their family as not ready at all for deployment and those who felt that their family coped poorly during deployment were also more likely to have a preference for leaving the Guard or Reserve.

CHAPTER NINE

What Are Guard and Reserve Families' Suggestions for Better Support?

In this chapter, we discuss what the service members and spouses we interviewed think the military could do to better support their families. Specifically, we posed an open-ended question also used in the 2006 Survey of Reserve Component Spouses: "How can the military provide better support for you and your family?" We maintain that understanding their suggestions, along with their expectations and opinions of the military, is an important aspect of our research. Since we used an open-ended question, spouses and service members indicated the suggestions that were most salient to them; it is possible if they had been presented with a list of ideas for improvement, they may have agreed that other actions would be beneficial as well.

As in the preceding chapters, we coded and analyzed the responses to this question to identify themes in the data and to determine the extent to which differences existed between spouses and service members, or within those two groups of individuals, in terms of what suggestions they emphasized. Figure 9.1 provides a breakdown of the proportions of service members and spouses that did or did not provide suggestions during the course of the interviews, and the figure shows that the two groups were very similar in their overall responses: 75 percent of each offered at least one suggestion; 17 percent indicated they did not have any suggestions because either they were already "doing fine" or they felt the military's current level and kind of support for families was already acceptable (i.e., that the military was also "doing

Figure 9.1
Summary of Service Member and Spouse Responses Regarding Suggestions

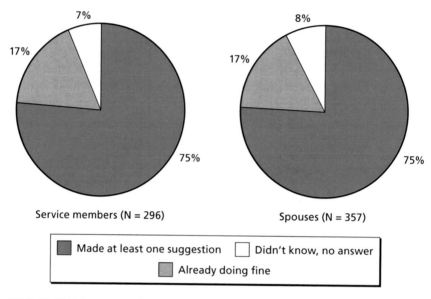

Service members (N = 296)   Spouses (N = 357)

| ■ Made at least one suggestion | □ Didn't know, no answer |
| ▨ Already doing fine | |

SOURCE: 2006 RAND Guard and Reserve Family Interviews.
NOTE: Percentages may not sum to 100 because of rounding.
RAND *MG645-9.1*

fine"); and 8 percent of spouses and 7 percent of service members either did not know what to say or opted not to answer the question.

Characteristics of those who did not offer suggestions because they were or the military was already doing fine tended to be ones associated with favorable outcomes in earlier sections of this monograph, such as higher family readiness, fewer problems related to deployment (in terms of mention), and better family coping. Table 9.1 provides a breakdown of all the characteristics for which a significant pattern was present among the spouses or the service members that we interviewed. For service members, those residing closer to their drill unit or the nearest military installation were more inclined to say they were already doing fine than were their counterparts living a greater distance from either location. Service members who had only experienced one OCONUS deployment since 9/11 also were more likely to decline

Table 9.1
Characteristics Associated with Stating "Already Doing Fine" in Lieu of Offering Suggestions

	Service Members (%)	Spouses (%)
Service member reserve component		
Army National Guard (N=102)		12
Army Reserve (N=89)		16
Air Force Reserve (N=83)		24
Marine Forces Reserve (N=83)		16
Distance from drill unit		
Less than 100 miles (N=218)	22	
100 or more miles (N=78)	5	
Distance from nearest military installation		
Less than 100 miles (N=247)	19	
100 or more miles (N=49)	8	
Amount of notice		
One month or less (N=157)		12
More than one month (N=189)		21
Perception of notice adequacy		
Adequate (N=133 service members; N=161 spouses)	24	23
Insufficient (N=55 service members; N=62 spouses)	4	6
Repeat OCONUS deployments since 9/11		
No (N=248)	19	
Yes (N=48)	8	
Family readiness		
Ready or very ready (N=192 service members; N=214 spouses)	21	22
Somewhat ready (N=43 service members; N=70 spouses)	12	10
Not at all ready (N=51 service members; N=55 spouses)	4	6

Table 9.1—Continued

	Service Members (%)	Spouses (%)
Problems		
Emotional or mental		
Cited problem (N=140)		10
Did not cite problem (N=217)		21
Household responsibilities		
Cited problem (N=144)		11
Did not cite problem (N=213)		20
Financial and legal		
Cited problem (N=42)	2	
Did not cite problem (N=254)	20	
Employment		
Cited problem (N=41)		2
Did not cite problem (N=316)		18
Education		
Cited problem (N=14)		0
Did not cite problem (N=343)		17
Marital		
Cited problem (N=39)		5
Did not cite problem (N=318)		18
No problems		
Cited "no problems" (N=87 service members; N=50 spouses)	23	46
Did not cite "no problems" (N=209 service members; N=307 spouses)	15	12
Positives		
Patriotism, pride, or civic responsibility		
Cited positive (N=44)	27	
Did not cite positive (N=252)	16	

Table 9.1—Continued

	Service Members (%)	Spouses (%)
No positives		
Cited "no positives" (N=60)	10	
Did not cite "no positives" (N=236)	19	
Family coping		
Coped well (N=186 service members; N=220 spouses)	19	24
Coped moderately (N=47 service members; N=71 spouses)	6	6
Coped poorly (N=25 service members; N=26 spouses)	12	0

SOURCE: 2006 RAND Guard and Reserve Family Interviews.

NOTES: Ns are provided for either service member or spouse, as denoted in the table. When data from both groups are shown, Ns are specified as service member or spouse. All percentages shown are statistically different from one another at p<0.10. Shading indicates a subset of population that is not significantly different from other subsets.

offering any suggestions for improvement. In addition, service members who stated that their family experienced financial and legal problems during deployment and those who claimed there were no positive aspects of their most recent activation were less likely to say they were already doing fine than were service members whose families did not experience this type of problem and those who mentioned at least one positive aspect, respectively. On the other hand, service members who cited patriotism, pride, or civic responsibility as a positive aspect of deployment were more likely to say their family was already doing fine; 27 percent of personnel who mentioned this positive aspect did not offer a suggestion, compared with 16 percent of personnel who did not identify an increased sense of patriotism, pride, or civic responsibility as a positive aspect.

Among the spouses, those married to Air Force reservists were more likely than those married to Army guardsmen to state they were already doing fine and not offer any suggestions for improvement. In addition, there was a relationship between the amount of notice received and declining to identify ways the military could better support spouses and their families: Spouses who received more notice were more likely to remark that the military did not need to do anything more than were those who received less notice. Conversely, spouses whose families experienced emotional, household, employment, education, or marital problems were less likely to do so. For example, only 10 percent of spouses who discussed an emotional or mental health problem during their interview later stated they were already doing fine, compared with 21 percent of spouses who did not mention a problem of this nature.

There were also several findings present in both the spouse and service member portions of the sample. Twenty-four percent of service members and 23 percent of spouses who deemed their notice adequate felt they were already doing fine, compared with just 4 percent of service members and 6 percent of spouses who believed their notice was insufficient. In addition, those who characterized their family as ready or very ready and those who felt that their family coped well during the deployment were more inclined to express this sentiment than those whose families were not ready or coped poorly. Finally, those who stated that their family did not experience any deployment-related problems also tended to decline to identify ways the military could provide better support for their family.

Turning our attention to those who did offer suggestions, spouses and service members cited a wide variety of ideas, listed below in the order of mention:

- Provide better or more information
- Make changes to benefits
- Improve family support programs and resources
- Make changes to reserve component operations
- Improve local resources for families
- Improve pay
- Improve reintegration support
- Improve notification
- Improve communication between the service member and family during deployment
- Connect reserve component spouses and families.

Suggestions offered by at least 10 percent of interviewees are shown in Figure 9.2. As the figure illustrates, none of these suggestions was provided by a majority of spouses or service members; the most frequently mentioned idea, to provide better or more information, was discussed by 24 percent of service members and 29 percent of spouses. In addition, spouses and service members resembled one another in the frequency with which they identified ideas, with a few statistically significant exceptions. Service members were more focused on changes to benefits; 21 percent of service members cited this, compared with 9 percent of spouses. Spouses, on the other hand, more frequently discussed the need for better communication with the service member during deployment and the usefulness of connecting spouses; 9 percent of spouses did so, versus only 2 percent of service members. In the sections that follow, we describe these opportunities for improvement, including any significant patterns among the service members

Figure 9.2
Suggestions Provided by Service Members and Spouses

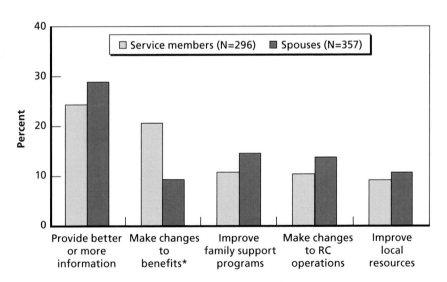

SOURCE: 2006 RAND Guard and Reserve Family Interviews.
*The service member and spouse percentages for this item were statistically different from one another at $p<0.10$.
RAND *MG645-9.2*

or spouses who tended to mention them. Since some of the ideas were proposed by relatively small numbers of spouses and service members, fewer patterns were evident, but those we did observe provide interesting insights. In addition, where appropriate, we consider the comments of the military family experts we interviewed, particularly those made in response to the question, "What do you think would be reserve families' biggest complaint about the support they're provided by the military?"

Provide Better or More Information

As mentioned above, "provide better or more information" was the most frequently cited recommendation for improvement by both service members and spouses. This is consistent with remarks made by the experts we interviewed; they tended to believe that families' biggest complaint about the support they were provided would be about the information they received—or what they did not receive—in a timely manner. As one DoD representative noted, "Their complaints will range from 'didn't get the information' to 'got too much information.'" (7: DoD military family expert) Another DoD interview participant stated:

> They don't know where to get help if they need help. They don't get the information about what's going on with their spouse from the unit. (4: DoD military family expert)

This second comment concisely highlights two main themes in the spouse and service member interviews: the need for better information regarding programs and services and the need for better information regarding the service member's deployment situation. Varied ideas were offered to improve the delivery of information about programs and services to families, as the following comments demonstrate:

> I think if the military can let us know what programs and other services are out there besides the Family Readiness Group before deployment, I think that would be more helpful. I guess an

example would be if we had known what programs or services were available at our closest military installation should we have needed them before it came down to actual deployment time. I think there are a lot of programs out there, but it isn't always easy to get that information. (8: Army Reserve, E-5's wife)

Increase awareness and have more classes. I know they have a lot of classes already. But have more classes during non-activation periods because when you're getting activated you're getting so much information all at once that it just goes in one ear and out the other. But if they do it one class at a time instead of having twenty classes on all these different things all at once, then it would make it better for you to absorb and before you get activated you're aware of all these people that can help you if you are ever in a bind. (65: Marine Forces Reserve, female E-5)

All the way through the deployment, continue to notify family members of any programs that are available, because it seems like at first, they have a whole lot of preparation for deployment and all this stuff for family members, and then once the deployment happens, it just stops. (481: Air Force Reserve, O-3)

While the first comment implies a failure to receive desired information at all, the latter two convey a different challenge—that so much information is distributed at the outset of a deployment, families cannot digest it all. Additional program and service-oriented comments related to the need for improved information on specific services, such as TRICARE, resolving errors in pay, and financial assistance.

The second type of information spouses especially felt needed improvement was about their guardsmen or reservist's service, in particular while he or she is deployed. As one spouse noted:

Getting information is very hard, and sometimes it's unstable. So a little bit more accurate information and a little bit of "know" for us is what we want—knowledge about my husband: where he is, how he's doing, what is his job. Also like I told you, with the coming home back and forth, the [details regarding] leaves and stuff haven't been accurate, so I've also missed out on chances to see him if it's a 24-hour pass because of the lack of time. He's cross-leveled, and I do live a far distance. (345: Army Reserve, O-2's wife)

Spouses also noted the frustration of waiting for accurate news about their service member, especially in light of general media reports or a specific injury, as the following comments illustrate:

Well, I think we should be able to hear from someone. I mean, sometimes I don't hear from him and I see the news. I don't know how they could, but it's just heartbreaking because you're at home and you hear about soldiers getting captured, or you hear about American soldiers getting killed over there. And here you don't have any support system. You can't find out if it's your husband or not. Days later they call, and all the time you have been worried. (316: Army Reserve, E-4's wife)

My husband, when he got hurt, I think they could change the way that they deal with a family member. The military—I know they have to have a certain type of way they talk to family members—but to contact the family member and tell them their loved one is hurt, and then you don't hear from them four, five days later, and you have to constantly call them, that's a problem for me. (322: Army Reserve, E-6's wife)

More accurate information about when the service member would be leaving and when he would be returning from the deployment was also requested.

A final set of information-related ideas pertained to more personal delivery of information to the family. Spouses and service members both mentioned the potential usefulness of status checks on the family during the deployment, which would serve a dual purpose of letting the family know the unit had not forgotten them and also providing information or other assistance appropriate to the family's current situation. As one spouse put it:

I just feel that if a family member is deployed, and they have a family left behind, I think that they should pick up the phone and see how that family is doing. And say, "Hey, do you need anything? We're just calling to check on you and your family to make sure that you guys are OK while your husband or your spouse is gone." (261: Air Force Reserve, E-5's wife)

Sometimes these comments directly referenced family support groups, and thus may also be considered in the context of the "improve family support" suggestions discussed below, but in other instances it was not clear whether the spouses and service members were focusing on family readiness–type groups as opposed to the military more broadly.

Some patterns were evident among the spouses and service members who identified information-related areas for improvement, and they are summarized in Table 9.2. Findings related to pay grade, reserve component, notice, and education problems were present for both service members and spouses. Specifically, among the service members we interviewed, mid-grade enlisted personnel were more likely to make this recommendation than were either junior enlisted personnel or junior officers. However, in the spouse portion of our interview sample, those married to junior officers were *more* inclined than those married to mid-grade enlisted personnel to suggest information-related improvements. With respect to reserve component, Army reservists were almost twice as likely to make this recommendation as Army guardsmen, and spouses married to Army guardsmen were less inclined to do so than with those married to Marine reservists. In addition, service members and spouses who felt the notice was insufficient were more likely to discuss this idea as well, as were those whose families experienced education-related problems during deployment. Finally, there were two unique findings present among only one portion of our interview sample. Service members who reported employment-related problems stemming from deployment were more inclined to recommend that the military provide more or otherwise better information than those who did not cite employment-related problems, and spouses who mentioned any deployment-related problems were more likely to make this suggestion than those who discussed at least one (of any type) during their interview.

Table 9.2
Characteristics Associated with Citing Provide Better or More Information as a Suggestion

	Service Members (%)	Spouses (%)
Service member pay grade		
E-1 to E-4 (N=69 service members; N=90 spouses)	19	26
E-5 to E-6 (N=146 service members; N=174 spouses)	32	26
O-1 to O-3 (N=81 service members; N=93 spouses)	15	37
Service member reserve component		
Army National Guard (N=104 service members; N=102 spouses)	18	22
Army Reserve (N=74 service members; N=89 spouses)	34	29
Air Force Reserve (N=60 service members; N=83 spouses)	25	30
Marine Forces Reserve (N=55 service members; N=83 spouses)	22	36
Perception of notice adequacy		
Adequate (N=133 service members; N=161 spouses)	23	22
Insufficient (N=50 service members; N=62 spouses)	40	42
Problems		
Employment		
Cited problem (N=61)	33	
Did not cite problem (N=235)	22	
Education		
Cited problem (N=41 service members; N=14 spouses)	39	50
Did not cite problem (N=255 service members; N=343 spouses)	22	28

Table 9.2—Continued

	Service Members (%)	Spouses (%)
No problems		
Cited "no problems" (N=50)		12
Did not cite "no problems" (N=307)		32

SOURCE: 2006 RAND Guard and Reserve Family Interviews.

NOTES: Ns are provided for either service member or spouse, as denoted in the table. When data from both groups are shown, Ns are specified as service member or spouse. All percentages shown are statistically different from one another at p<0.10. Shading indicates a subset of population that is not significantly different from other subsets.

Make Changes to Benefits

Making changes to benefits was the second most frequently mentioned suggestion overall, but as discussed above, service members cited this idea more frequently than did spouses; 21 percent of service members offered recommendations of this nature, compared with just 9 percent of spouses. Spouses and service members both tended to refer to changes to TRICARE. Interestingly, while some spoke positively of TRICARE and advocated expanding its availability to reserve component families, others focused more on improving TRICARE inadequacies, as the following remarks demonstrate:

> TRICARE could be easier to get through. All of the teleprompting and the non-real people in TRICARE is a little tough. (139: Air Force Reserve, E-5's wife)

> Well, the big thing that we felt was lacking was TRICARE, frankly. It's very much a government bureaucracy; it's not always easy to navigate to get what you need when you need it. It seems to me with something so important as having medical care while you're away for your family, I think it could be more user-friendly and easier for family members to take advantage of. Every time we had a question, you had to call the 1-800 number, and you'd have to navigate through their different menus to get answers. It wasn't always easy to speak to someone to get an answer. With TRI-

CARE, a lot of physicians don't accept TRICARE, so it makes
your choice of providers somewhat limited; you can't always go
to who you'd want to go to because they don't take TRICARE.
That's really the big thing that I had a problem with. (178: Marine
Forces Reserve, O-3)

In addition to discussing TRICARE, spouses mentioned improvements to financial benefits and to programs related to children. Specifically, a small number of spouses discussed financial benefits, such
as paying for their education, lower interest rates, and improved access
to emergency relief funds. Similarly small numbers recommended
child care improvements and additional programs for children. Service
members seldom mentioned those types of benefit changes; instead, a
handful of personnel discussed changes to retirement eligibility and
benefits parity with the Active Component.

Patterns associated with a tendency to recommend changes to benefits are shown in Table 9.3. Most of them were present only in the
service member portion of the sample, perhaps due in part to the larger
proportion of service members who suggested actions of this nature.
Specifically, male service members and junior officers were more likely
to cite suggestions of this nature, as were those who lived farther away
from their drill unit. Conversely, service members who cited either
patriotism/pride/civic responsibility or family independence/confidence/
resilience as positive aspects of their activation were less inclined to recommend changes to benefits than those who did not mention these
positives. Among the spouses, those who lived farther from the nearest
military installation were more inclined to discuss the need for improvements to benefits. Turning our attention toward common characteristics, both amount of notice and health care problems during deployment were associated with how frequently both service members and
spouses stated the military needed to make changes to benefits. For service members, those receiving less notice tended to offer this suggestion,
but for spouses, it was more often mentioned by those who received
more notice. Lastly, those who mentioned health care–related problems
stemming from deployment were more inclined to recommend changes

Table 9.3
Characteristics Associated with Citing Make Changes to Benefits as a Suggestion

	Service Members (%)	Spouses (%)
Overall percentages citing make changes to benefits (N=296 service members; N=357 spouses)	21	9
Gender		
Male (N=270)	22	
Female (N=26)	8	
Service member pay grade		
E-1 to E-4 (N=69)	16	
E-5 to E-6 (N=146)	19	
O-1 to O-3 (N=81)	27	
Distance from drill unit		
Less than 100 miles (N=218)	17	
100 or more miles (N=78)	31	
Distance from nearest military installation		
Less than 25 miles (N=148)		6
25 or more miles (N=175)		12
Amount of notice		
One month or less (N=156 service members; N=157 spouses)	25	6
More than one month (N=140 service members; N=189 spouses)	16	12
Problems		
Health care		
Cited problem (N=35 service members; N=41 spouses)	49	42
Did not cite problem (N=261 service members; N=316 spouses)	17	5
Positives		
Patriotism, pride, or civic responsibility		
Cited positive (N=44)	9	
Did not cite positive (N=252)	23	

Table 9.3—Continued

	Service Members (%)	Spouses (%)
Spouse/child independence, confidence, or resilience		
Cited positive (N=40)	8	
Did not cite positive (N=256)	23	

SOURCE: 2006 RAND Guard and Reserve Family Interviews.

NOTES: Ns are provided for either service member or spouse, as denoted in the table. When data from both groups are shown, Ns are specified as service member or spouse. All percentages shown are statistically different from one another at p<0.10. Shading indicates a subset of population that is not significantly different from other subsets.

to benefits than were service members and spouses whose families did not encounter this challenge.

Improve Family Support Programs and Resources

About 13 percent of our interview participants suggested that improving family support programs and resources would help better support reserve families. Comments included references to both Family Readiness Groups and Key Volunteer Networks in particular as well as to military-sponsored family support efforts more broadly. Remarks varied greatly within this common theme. Some spouses and service members discussed the need to tailor support—to make it more reserve-specific, for example—while others focused on subject matter such as readiness. Still others discussed the types of people who especially needed more support, such as those far away from a military installation or those cross-leveled into a different unit:

> [I]t's a very lonely feeling to send your husband away for a year and to feel like you don't have any support. Again, it might be because of my situation where my husband was deployed with a unit on the other side of the country. Had he been deployed with people right here in my hometown, it could have been a completely different situation, but officers are being tasked out to

go with units in other locations a lot more these days and I don't think the military's paying enough attention to those situations in terms of the families that are left behind. (767: Army Reserve, O-2's wife)

Organize the Family Readiness Group a little better. Living away from the base as far away as we do, I still feel very isolated even though we have the Family Readiness Group. I don't feel like they have been doing a lot to help other than giving me information about the guys. Supposedly the goal of the program is to make sure everybody is OK while they are gone, and I don't feel like they have done that well enough. (149: Marine Forces Reserve, E-5's wife)

Well, we're probably a little different since we didn't have a unit nearby. I didn't deploy as a whole unit. So, really, there wasn't any unit support because of that. And I think probably they should look to provide some kind of support or information either via a website or some larger organization because the family support groups are great if you deploy with a unit. But when you don't, then those people just fall through the cracks. The military really didn't assist my family at all because of that, I think. I know the military can do a good job. I've deployed with units before, and it's a big difference. (211: Army Reserve, O-3)

As the last comment suggests, some interview participants viewed Internet-based resources as one way to bridge the gap between families and the military. One spouse suggested, "Start a secured [Web] site where families can establish a Yahoo group,[1] where they can list and post and put files up" (383: Army Reserve, E-6's wife).

Lastly, spouses and service members felt that those who run family support efforts should be better qualified or receive training before providing family support. As one service member put it, "Family support

---

[1]   According to Yahoo! Inc., "Yahoo! Groups is a free service that allows you to bring together family, friends, and associates through a web site and email group. Yahoo! Groups offer a convenient way to connect with others who share the same interests and ideas" (Yahoo! Inc., no date).

groups, Family Readiness Groups—some of them are not well prepared for some of these deployments, and they need to be more educated" (145: Army National Guard, E-5). A small number of interviewees also suggested reducing the dependence on volunteers for this type of support, as follows:

> I think one thing is that there are the family support groups in this area, but they are all run by military wives that are under stress themselves. They're already having a hard time with their families. I think they should be run by somebody that isn't in this situation. I think it should be a full-time job, not just a volunteer position. Their husbands are already gone, and now they're expected to do all this work, emailing all these people, coordinating all these picnics and events, child care and all this stuff. That's a lot for them to take on. (642: Army National Guard, E-4's wife)

> [My suggestion is] to have someone who is active duty or federally employed or a federally employed contractor who is in-charge of family support. Not a volunteer, not a spouse of a chain of the command. Someone who is being paid by the Department of Defense to oversee each unit's family support. (66: Army Reserve, E-4)

The military family experts we interviewed also suggested potential problems that stem from a reliance on volunteers. For instance, one mentioned the danger of "volunteer burnout" (18: non-DoD military family expert), while another believed that military-sponsored family support "depends too heavily on volunteer support" (9: DoD military family expert).

Table 9.4 shows the characteristics that were associated with the frequency of mention for this recommendation. Within the service member portion of the sample, junior officers and those living farther away from their drill unit tended to identify this as an area for improvement more frequently. In addition, service members who indicated that their family coped well during their deployment were less likely to offer this suggestion than those whose families coped either moderately or poorly. A look at the spouse-only patterns reveals that those married

Table 9.4
Characteristics Associated with Citing Improve Family Support Programs and Resources as a Suggestion

	Service Members (%)	Spouses (%)
Service member pay grade		
E-1 to E-4 (N=69 service members; N=83 spouses)	13	
E-5 to E-6 (N=146 service members; N=163 spouses)	8	
O-1 to O-3 (N=81 service members; N=84 spouses)	15	
Service member reserve component		
Army National Guard (N=104 service members; N=102 spouses)	8	17
Army Reserve (N=74 service members; N=89 spouses)	22	19
Air Force Reserve (N=60 service members; N=83 spouses)	9	8
Marine Forces Reserve (N=58 service members; N=83 spouses)	11	13
Service member prior active duty		
Yes (N=206)		12
No (N=151)		19
Distance from drill unit		
Less than 100 miles (N=218)	9	
100 or more miles (N=78)	17	
Perception of notice adequacy		
Adequate (N=161)		12
Insufficient (N=62)		24
Deployment length		
Less than one year (N=180)		11
One year or more (N=169)		19
Problems		
Emotional or mental		
Cited problem (N=140)		21
Did not cite problem (N=217)		10

Table 9.4—Continued

	Service Members (%)	Spouses (%)
Children's issues		
Cited problem (N=94)		20
Did not cite problem (N=263)		13
Financial and legal		
Cited problem (N=58)		22
Did not cite problem (N=299)		13
Marital		
Cited problem (N=36 service members; N=39 spouse)	25	28
Did not cite problem (N= 260 service members; N=318 spouses)	9	13
No problems		
Cited "no problems" (N=87 service members; N=50 spouses)	6	4
Did not cite "no problems" (N=209 service members; N=307 spouses)	13	16
Family coping		
Coped well (N=186)	9	
Coped moderately (N=47)	19	
Coped poorly (N=25)	16	

SOURCE: 2006 RAND Guard and Reserve Family Interviews.

NOTES: Ns are provided for either service member or spouse, as denoted in the table. When data from both groups are shown, Ns are specified as service member or spouse. All percentages shown are statistically different from one another at p<0.10. Shading indicates a subset of population that is not significantly different from other subsets. For reserve component comparisons in the service member group, the Army Reserve is significantly different from the three other components. For reserve component comparisons in the spouse group, the Air Force Reserve is significantly different from the Army Reserve and the Army National Guard. Other reserve component comparisons are not significantly different.

to service members without prior active duty experience, those who regarded the notice received as insufficient, those with longer deployments, and those whose families experienced emotional, child-related, or financial or legal problems during deployment were more inclined to state that the military could improve family support. For example, 21 percent of spouses who mentioned that their family had an emotional or mental problem also recommended improvements to family support programs, compared with just 10 percent of spouses who did not discuss this type of problem during their interview. For both spouses and service members, those affiliated with the Air Force Reserve were less inclined to make this suggestion; Army guardsmen also less frequently recommended this action than did Army reservists and Marine reservists. In addition, interviewees who discussed marital problems stemming from deployment tended to assert a need for family support program improvements. Lastly, those who asserted their family did not experience any deployment-related problems were less likely to recommend this change.

Make Changes to Reserve Component Operations

Twelve percent of interviewees suggested changing reserve component operations in ways they believe would help them and their family. Many of their ideas pertained to changing the nature of deployments, particularly in terms of their length and frequency. The following comments are typical of this theme:

> Every soldier has to make sacrifices, but it's not fair that several people are on their second or third deployment and there are others that haven't even deployed at all. So, shorter deployments and everyone in uniform should be required to go or get out. You don't deserve to wear the uniform if you're not going to go. It's not fair. (94: Army Reserve, female O-3)

> Be cognizant of how many times, close together, and the length that you activate your personnel, because they can't recover. Mostly I would say Army because my brother-in-law is in the Army, and they activate him and keep him just right under the time frame where they can't activate him for three years, and then they send him home. And then in four months they activate him again. So, we're not typical. I think the Air Force does a really good job of

everybody doing their share. But most of the time they're not front line personnel either. (362: Air Force Reserve, E-7's wife)

Probably one of my only quirks [about] being reserve component or National Guard is the amount of time for activation and deployment. Most of us have careers and families already established, I think that the time spent over there is a little bit lengthy. They do a good job of supporting us, which like I said, is definitely a good thing. I just wish that the deployment time would be broken up into shorter deployments. (214: Army National Guard, O-1)

For one, they don't have to deploy a family member for 18 months. That's pretty ridiculous. I understand the whole deal and that stuff, but you don't see regular Army and Navy leaving their families for 18 months at a time. (79: Army National Guard, E-6's husband)

As the last comment, made by an Army National Guard spouse, indicates, the long deployments experienced by the Army National Guard in particular (recall from Chapter Two their average deployment length was the highest in our sample) likely motivated this recommendation. In addition, a small number of spouses from all four components in our study also suggested not deploying reserve component personnel OCONUS, as these remarks illustrate:

I just think that if they want to go to another country to fight a war they should use people who are in the military and leave the actual National Guard and reserve home, which is what they were meant for. (59: Army National Guard, E-6's wife)

Stop sending our husbands overseas. Is that an answer [to the question]? If Texas gets attacked, then that's fine because we can go defend Texas. (780: Marine Forces Reserve, O-3's wife)

This may stem from some of the spouses' surprise at an OCONUS deployment. As one of our experts explained:

For guard families, before 9/11, activations were few and far between—a big gap between Vietnam and Desert Storm. We haven't done it [large-scale deployments] for two generations, so guard families didn't realize they were possibly going to be activated and deployed to a warfighting environment. (6: DoD military family expert)

Other, less frequently mentioned changes to reserve component operations include improving the promotion process and reducing the practice of cross-leveling personnel. Finally, a few spouses simply requested, "Bring him home."

Table 9.5 shows how comments of this nature tended to be distributed among the service members and spouses we interviewed. The bulk of significant patterns was present only in the spouse portion of the interview sample. Specifically, male spouses, those married longer, and those living closer to the nearest military installation were more likely to recommend changes to reserve component operations than were female spouses, newlyweds, and those living farther from the nearest installation, respectively. In addition, spouses who reported problems related to household responsibilities, children's issues, or marital issues tended to offer this suggestion more frequently, as did spouses who asserted there were no positive aspects to their service member's activation. On the other hand, spouses who indicated that their family coped well with deployment were less inclined to express a need for changes to reserve component operations than those whose families coped moderately or poorly.

Findings related to age, pay grade, employment problems, and the absence of problems were apparent among the service members we interviewed. Older service members and junior officers were both more likely to request changes to reserve component operations. In addition, 20 percent of service members who discussed employment problems expressed this sentiment, compared with just 8 percent of service members who did not cite this type of problem. In a related vein, service members who said that their family did not encounter any kind of deployment-related problem were less inclined to make this recommendation.

Lastly, one common finding was present for both service members and spouses; consistent with the preceding observations about the Army National Guard's long deployment length, those affiliated

Table 9.5
Characteristics Associated with Citing Make Changes to Reserve Component Operations as a Suggestion

	Service Members (%)	Spouses (%)
Age		
25 or less (N=37)	3	
26 or more (N=258)	12	
Gender		
Male (N=12)		42
Female (N=345)		13
Marriage length		
2 years or less (N=71)		4
3 years or more (N=286)		16
Service member pay grade		
E-1 to E-4 (N=69)	9	
E-5 to E-6 (N=146)	6	
O-1 to O-3 (N=81)	20	
Service member reserve component		
Army National Guard (N=104 service members; N=102 spouses)	14	22
Army Reserve (N=74 service members; N=89 spouses)	14	15
Air Force Reserve (N=60 service members; N=83 spouses)	7	7
Marine Forces Reserve (N=58 service members; N=83 spouses)	5	10
Distance from nearest military installation		
Less than 25 miles (N=148)		18
25 or more miles (N=175)		11
Problems		
Household responsibilities		
Cited problem (N=144)		18
Did not cite problem (N=213)		11

Table 9.5—Continued

	Service Members (%)	Spouses (%)
Children's issues		
Cited problem (N=94)		21
Did not cite problem (N=263)		11
Employment		
Cited problem (N=61)	20	
Did not cite problem (N=235)	8	
Marital		
Cited problem (N=39)		26
Did not cite problem (N=318)		12
No problems		
Cited "no problems" (N=87)	6	
Did not cite "no problems" (N=209)	12	
Positives		
No positives		
Cited "no positives" (N=48)		29
Did not cite "no positives" (N=309)		11
Family coping		
Coped well (N=220)		9
Coped moderately (N=71)		21
Coped poorly (N=26)		27

SOURCE: 2006 RAND Guard and Reserve Family Interviews.

NOTES: Ns are provided for either service member or spouse, as denoted in the table. When data from both groups are shown, Ns are specified as service member or spouse. All percentages shown are statistically different from one another at p<0.10. Shading indicates a subset of population that is not significantly different from other subsets. For pay grade comparisons in the service member group, the O-1 to O-3 category is significantly different from the E-1 to E-4 and E-5 to E-6 categories. The other pay grade comparison is not significantly different. For reserve component comparisons in the spouse group, the Army National Guard is significantly different from the Air Force Reserve and the Marine Forces Reserve. The other reserve component comparisons are not significantly different.

with the Army National Guard were more likely to suggest changes to reserve component operations. Fourteen percent of Army guardsmen made this recommendation, compared with just five percent of Marine reservists. Similarly, 22 percent of Army National Guard spouses did so as well, compared with 10 percent of Marine Forces Reserve spouses and 7 percent of Air Force Reserve spouses.

Improve Local Resources for Families

Ten percent of interviewees identified a need for more support resources close to their homes, and the proportions were comparable for the service member and spouse portions of the sample. While some of the comments referred to the need for local family support from the units themselves, more remarks pertained to a need for local resources, particularly those readily available at a military installation for families of deployed active component personnel. These sentiments are conveyed in the following remarks:

> Here we don't have a lot of the resources that you would have if you were closer to a major installation, like commissaries and stuff of that nature. It's difficult sometimes to find a doctor or dentist that takes the TRICARE. That's just an issue of being where I'm at geographically, 'cause I know the Army has multiple resources at major installations, but we just don't have access to a lot of those. Well, I think if they were to set up more of a co-op with civilian resources, maybe that would help. (103: Army National Guard, E-4)

> They need to provide better resources for things we *really* need like help with the yard work. Things that our spouses usually do, we need help with. In my area, Milwaukee is not that small a city and there is only one place where they have that day care thing and it is in the ghetto. So better resources, more options for daycare, help with yard work. Whatever we need—mechanics, plumbers, things like that. Things that happen around the house I don't know how to fix—things in the yard, things that most of

the time a man is the one who takes care of. (548: Army Reserve, E-4's wife)

The last comment also suggested that local resources were viewed as important because they help offset the absence of the service member, who may have had primary responsibility for a wide array of household chores.

As with the last recommendation, only one characteristic served as the basis for significant differences in the responses of both service members and spouses. In this case, that factor was family readiness. However, the direction of the pattern differed for the two portions of our interview sample. As shown in Table 9.6, service members who characterized their family as not ready at all were more likely to suggest improving local resources than those who felt that their family was somewhat ready or better for deployment. Among the spouses we interviewed, however, those who indicated their family was ready or very ready were most inclined to offer this recommendation.

While there was only one common characteristic associated with mentioning this recommendation, many bases for significant differences were present among only the service members or only the spouses that we interviewed. With respect to service members, female service members, Army reservists, those living farther away from their drill unit, those living farther away from the nearest military installation, and those who thought the notice they received was insufficient all were more inclined to state a need for better local family resources. Deployment-related problems also helped to account for differences in who tended to make this recommendation: Service members who described emotional, financial, or health care problems stemming from their deployment were more inclined to advocate better local resources for their family than their counterparts whose families did not face such challenges. Conversely, service members who reported that their family did not experience problems of any type during their deployment and those who viewed greater family closeness as a positive consequence of their deployment were less likely to offer this suggestion.

Turning our attention to the spouses we interviewed, a similarly large number of different factors help to characterize the types

Table 9.6
Characteristics Associated with Citing Improve Local Resources for Families as a Suggestion

	Service Members (%)	Spouses (%)
Age		
25 or less (N=55)		2
26 or more (N=302)		12
Gender		
Male (N=270)	0	
Female (N=26)	10	
Marriage length		
2 years or less (N=71)		3
3 years or more (N=286)		13
Parental status		
Has children (N=269)		12
No children (N=88)		6
College degree		
Yes (N=195)		13
No (N=162)		7
Service member pay grade		
E-1 to E-4 (N=90)		6
E-5 to E-6 (N=174)		15
O-1 to O-3 (N=93)		8
Service member reserve component		
Army National Guard (N=104 service members; N=102 spouses)	7	
Army Reserve (N=74 service members; N=89 spouses)	16	
Air Force Reserve (N=60 service members; N=83 spouses)	7	
Marine Forces Reserve (N=58 service members; N=83 spouses)	7	

Table 9.6—Continued

	Service Members (%)	Spouses (%)
Distance from drill unit		
Less than 25 miles (N=83)	4	
25 or more miles (N=213)	11	
Distance from nearest military installation		
Less than 25 miles (N=120)	3	
25 or more miles (N=176)	14	
Less than 100 miles (N=247)	7	
100 or more miles (N=49)	22	
Perception of notice adequacy		
Adequate (N=133 service members; N=148 spouses)	6	
Insufficient (N=55 service members; N=59 spouses)	16	
Repeat OCONUS deployments since 9/11		
No (N=225)		8
Yes (N=132)		14
Family readiness		
Ready or very ready (N=192 service members; N=214 spouses)	7	13
Somewhat ready (N=43 service members; N=70 spouses)	7	4
Not at all ready (N=51 service members; N=55 spouses)	20	7
Problems		
Emotional or mental		
Cited problem (N=78)	14	
Did not cite problem (N=218)	7	
Financial and legal		
Cited problem (N=42)	17	
Did not cite problem (N=254)	8	

Table 9.6—Continued

	Service Members (%)	Spouses (%)
Health care		
Cited problem (N=35)	17	
Did not cite problem (N=261)	8	
No problems		
Cited "no problems" (N=87)	5	
Did not cite "no problems" (N=209)	11	
Positives		
Family closeness		
Cited positive (N=60)	3	
Did not cite positive (N=234)	11	

SOURCE: 2006 RAND Guard and Reserve Family Interviews.

NOTES: Ns are provided for either service member or spouse, as denoted in the table. When data from both groups are shown, Ns are specified as service member or spouse. All percentages shown are statistically different from one another at p<0.10. Shading indicates a subset of population that is not significantly different from other subsets. For pay grade comparisons in the spouse groups, the E-5 to E-6 category is significantly different from the E-1 to E-4 and O-1 to O-3 categories. The other pay grade comparison is not significantly different. For reserve component comparisons in the service member group, the Army Reserve is significantly different from the Army National Guard and the Air Force Reserve. Other reserve component comparisons are not significantly different.

of spouses who desired improvements to local family resources. More mature spouses, as indicated by their older age or longer marriage length, were more inclined to recommend this course of action, as were spouses with a college degree. In addition, spouses married to mid-grade enlisted personnel were more likely than others to recommend that the military improve its local resources for families. Lastly, spouses who were parents and those who experienced multiple OCONUS deployments since 9/11 also tended to propose improvements to local family resources.

Additional Suggestions

Small numbers of spouses and service members mentioned the remaining suggestions: improve pay, in terms of both its amount and timeliness; improve reintegration support; improve notification; improve communication between the service member and his family during deployment; and connect spouses. Each of these was identified by less than 7 percent of our overall interview sample. As noted earlier, we used an open-ended question to elicit suggestions for improvement, and did not present a list of possible actions. Thus, some of these less frequently cited ideas may still be viewed as helpful by spouses and service members; they were simply less salient to those we interviewed. In the case of reintegration support, this also may have been mentioned less frequently because many of the spouses we interviewed were still in the midst of their service member's deployment. They had not yet reached the post-deployment phase and its potential challenges.

Of particular note here are the recommendations to improve communication with the service member and to connect spouses. In both cases, those mentioning these ideas were disproportionately spouses.[2] Spouses felt communication with their service member was too infrequent, too costly, or simply too difficult:

> [Provide] more means of communication between the families and the spouse while they're deployed. Like some places have Internet, some places don't. Some places, they can call home anytime they want to. Some places, very limited phone use. (384: Army Reserve, E-6's wife)

> More lines of communication with your spouse. When he was in Iraq, he had to stand in a long line to use the phone. Most of our conversation was through email. Most of the time the system was down, especially if someone was killed in the group. They

---

[2]   Improve communication with the service member was recommended by 2 percent of service members and 9 percent of spouses. "Connect spouses" was not suggested by any service members, and it was mentioned by 6 percent of spouses. In both cases, differences between service member and spouse response frequencies were statistically significant at $p<0.10$.

shut things down when someone was killed. (625: Army National Guard, E-4's wife)

Many of the comments also described how the telephone cards many service members were provided either did not work or were valid for far fewer minutes than advertised. As one spouse put it, "When they [service members] need to call home, I think that they should supply some type of way so they can call home without using those calling cards that eat up the minutes real fast" (73: Army National Guard, E-5's wife).

The idea to connect spouses with one another was not mentioned by any service members. Instead, spouses exclusively described this as a way to help guard and reserve families. These spouses felt that one way to reduce or even avoid the feelings of isolation and loneliness that arise during deployment, especially for those away from military support, would be to connect spouses and families in the same situation. The following comments illustrate this theme:

> For myself, I feel like you're alone. I don't feel like you have any support from the military base. When my husband was active duty, you felt like you were a part of a community, but with my husband in the reserves, I feel like he goes to the reserves and comes home. There's no sense of community feeling. I don't know any of the other families. You don't have anybody to share what you're going through with any of the other families. I feel they could provide a better sense of community within the reserve community. (90: Air Force Reserve, E-5's wife)

Even if the families did not live near other families in the same reserve component, much less from the same unit, spouses felt there was still a shared experience that could help foster a much-needed sense of community:

> It's hard with reserves because we're so far apart. The best thing for families dealing with this is to have each other to rely on and a lot of reservists in my husband's unit especially are from Maine to Connecticut so I [don't] have, that kind of community feeling, when you can talk to someone that's going through it. . . . You

know what I mean? They may not be in your group, they may not be in your battalion, they may not be in the Marines, they may be in the Army. The closest woman that I know that I would like to talk to is well over an hour or two hours away, and when you're a single parent working full-time and doing laundry, it's hard to make that [trip], you know, to pay for that gas. But I bet there's people near me that are maybe living here, but they're in the Army. If we were full-time military we'd be on a base. I think that's what's missing. It's really difficult. (153: Marine Forces Reserve, E-5's wife)

Discussion

Service members and spouses were asked via an open-ended question to identify ways in which the military could better support guard and reserve families. The frequency with which these suggestions were cited provides a sense of how salient they were to the spouses and service members we interviewed, yet the proportion of spouses and service members who might view each as helpful may be even higher. Seventy-five percent of the interview participants offered at least one suggestion, while 17 percent did not offer suggestions because they either were already doing fine or thought the current level of support was acceptable. The spouses and service members in our study did not present one predominant suggestion to help reserve families; instead they offered a variety of ideas.

The most frequently mentioned suggestion was to provide better or more information. Spouses and service members wanted information about programs and services, as well as about aspects of the service member's deployment, delivered in a timely manner. They also recommended changes to benefits, most notably TRICARE, and suggested ways to improve the provision of family support programs and resources. Changes to reserve component operations were also cited by both spouses and service members, with most of these comments focusing on aspects of deployment, such as its length, frequency, or location. Providing local resources to assist spouses in handling tasks typically the responsibility of the service member was also presented as a way to help reserve families. In the same vein, service members and spouses also encouraged the military to find ways to offer resources locally to families who cannot readily access them at military installations. Smaller numbers of service members and spouses also suggested improvements to pay (both timing and amount), reintegration support, notification, and communication with the service members. Service members and spouses tended to offer suggestions in similar proportions, but military personnel were more focused on changes to benefits than were spouses. On the other hand, spouses were more inclined to suggest ways to improve communication with the service member during deployment, and only spouses mentioned the need to connect spouses and families to combat feelings of isolation and foster a sense of community not also present within the guard and reserve.

Table 9.7 summarizes the factors we examined in our analysis of the suggestions for how the military can better support families. Specifically, both patterns associated with a spouse or service member's tendency to offer a specific suggestion and those for which no such relationship was present are shown for the suggestions offered by at least one-tenth of our sample, as well as for those who indicated they were already doing fine.

Gender and age both had some relationship to offering specific suggestions. The female service members we interviewed tended to recommended improving these resources, while male service members were more inclined to suggest changes to benefits. In an additional gender-oriented finding, the husbands in the spouse portion of our sample were more likely to recommend changes to reserve component operations than were the wives we interviewed. With respect to age, older service members were more inclined to request changes to reserve component operations, and older spouses focused on improving local resources for families. Response patterns based on other indicators of maturity—marriage length and college degree—were present as well: Spouses married longer more frequently requested reserve component operational changes and better local resources than did newlyweds, and those with a college degree were more inclined to discuss improving local resources than spouses without this credential. Spouses with children also tended to offer this suggestion as well.

Table 9.7
Summary of Factors Related to Suggestions for Better Family Support

	Already Doing Fine	Provide Better or More Information	Make Changes To Benefits	Improve Family Support Programs And Resources	Make Changes to Reserve Component Operations	Improve Local Resources For Families
Individual and Situational Characteristics						
Age					SM	SP
Gender			SM		SP	SM
Marriage length					SP	SP
Parental status						SP
College degree						SP
Service member pay grade		SM, SP	SM	SM	SM	SP
Service member reserve component	SP	SM, SP		SM, SP	SM, SP	SM
Service member prior active duty				SP		
Distance from drill unit	SM		SM	SM		SM
Distance from nearest military installation	SM		SP		SP	SM
Amount of notice	SP		SM, SP			
Deployment length				SP		
Perception of notice adequacy	SM, SP	SM, SP		SP		SM
Repeat OCONUS deployments	SM					SP
Family readiness	SM, SP					SM, SP

Table 9.7—Continued

	Already Doing Fine	Provide Better or More Information	Make Changes To Benefits	Improve Family Support Programs And Resources	Make Changes to Reserve Component Operations	Improve Local Resources For Families
Problems						
Emotional or mental	SP			SP		SM
Household responsibilities	SP				SP	
Children's issues				SP	SP	
Financial and legal	SM			SP		SM
Employment	SP	SM			SM	
Education	SP	SM, SP			SM	
Marital	SP			SM, SP	SP	
Health care			SM, SP			SM
No problems	SM, SP	SP		SM, SP	SM	SM
Positives						
Family closeness						SM
Patriotism, pride, or civic responsibility	SM		SM			
Spouse/child independence			SM			
No positives	SM				SP	
Family coping	SM, SP			SM	SP	

SOURCE: 2006 RAND Guard and Reserve Family Interviews.

NOTES: All relationships listed are statistically significant at $p<0.10$. SM = Finding present in the service member portion of the sample (N=296). SP = Finding present in the spouse portion of the sample (N=357).

Turning our attention to military-related characteristics, findings related to service member pay grade and reserve component were apparent. Mid-grade enlisted personnel were more likely to suggest providing more or better information, while junior officers advocated changes to benefits, improvements in family support resources, and modifications to reserve component operations. Spouses of junior officers were more inclined to request better or more information than were those married to mid-grade enlisted personnel, while spouses of mid-grade enlisted personnel were more likely than those married to other personnel to suggest improvements to local family resources. In addition, Army reservists were more likely than Army guardsmen to recommend better or more information; more likely than either Army guardsmen or Air Force reservists to request improved local resources for families; and more likely than service members from the other reserve components studied to mention a need for improved family support programs and resources. Army guardsmen were more likely than Marine reservists to recommend changes to reserve component operations, and similarly, spouses of Army guardsmen were more inclined than spouses of either Marine reservists or Air Force reservists to express this sentiment. On the other hand, spouses married to Army guardsmen were less likely to cite a need for better or more information or to state they were already fine. Spouses married to service members with prior active duty experience were also less inclined to recommend improvements to family support programs.

Distance from one's drill unit and distance from the nearest military installation were associated both with a greater likelihood to state that the current situation was already fine and to offer specific suggestions; service members living closer to their drill unit tended to believe things were already fine, while service members living farther away from their drill unit were more likely to recommend changes to benefits, improvements to family support, and improvements to local resources for families. Likewise, service members living closer to the nearest military installation were more inclined to assert their family was doing fine (i.e., no actions necessary), while those living farther away perceived a need to improve local resources.

With respect to families' deployment experience, service members who had completed only one OCONUS deployment since 9/11 were more likely to believe the current situation was fine, and spouses who experienced only one such deployment were less inclined to suggest improvements to local resources for families. Deployment length was also related to suggestions from spouses; specifically, spouses whose service members experienced a deployment of one year or longer were more inclined to ask for improved family support programs and resources. In addition, we noted mixed findings related to the amount of notice. For service members, less notice was associated with more frequently suggesting a change in benefits, whereas for spouses, less notice was associated with less frequently doing so. Perhaps spouses who received minimal notice were less concerned with benefit changes than they were with other improvements. In addition, more notice was associated with a greater tendency to indicate that the current situation was already fine, as were perceptions that the amount of notice was adequate and that family readiness was high. Perceptions of notice adequacy were also related to several suggestions: Those who felt that their notice was insufficient were more inclined to suggest the need for better or more information, improvements to family support, or improved local resources for families. Family readiness was also related to citing a need for better local resources; service members whose families were not ready at all were more likely to recommend actions of this nature, while spouses whose families were ready or very ready were more likely to make this recommendation. The reason for these opposite results is not clear; it may be due to different experiences encountered by service member and spouse families during deployment, or it could be an example of spouses and service members' tendency to perceive situations differently.

Problems stemming from deployment were often associated with more frequent references to ways the military could help families, especially with respect to such actions as providing better or more information, improving family support programs, and altering reserve component operations. Spouses whose families experienced emotional or mental problems were more inclined to recommend improvements to family support programs and resources than those who did not discuss problems of this nature, and service members who cited emotional problems tended to request improvements to local resources for families. Among the spouses we interviewed, those who described household problems tended to request changes to reserve component operations more frequently, and those who reported children's issues had a greater likelihood of recommending improvements to family support programs as well as modifications to reserve component operations. Those who cited financial and legal difficulties tended to suggest improvements to family support programs and to local resources for families. Those who encountered employment-related problems asked the military to provide better or more information and to make changes to reserve component operations; and those with education-oriented challenges tended to focus on information-related improvements as well. Marital difficulties were associated with more frequently requesting family support program improvements and changes to reserve component operations, and health care problems were related to more frequent recommendations to make changes to benefits and to improve local resources for families—two suggestions with likely implications for TRICARE.

Conversely, those who reported that their family did not experience problems stemming from deployment were less likely to recommend providing better or more information, improving family support programs, making changes to reserve component operations, or improving local resources for families. In a similar vein, those who did not offer any recommendations because their family was already fine were also less inclined to mention many of these problems and more inclined to assert that their family did not experience any problems as a result of deployment.

In addition, several of the positives were related to a reduced tendency to make recommendations, especially on the part of service members. Service members who discussed family closeness as a positive aspect of their deployment were less likely to state a need for better local family resources, and those who viewed increased patriotism, pride, or civic responsibility or increased spouse or child independence, confidence, or resilience as positives were less inclined to request changes to benefits. Those who cited a heightened sense of patriotism, pride, or civic responsibility also tended to decline to offer any recommendations. On the other hand, service members who said there were no positive aspects of their deployment were less inclined to say that their family was already fine, and spouses who perceived no positives more frequently discussed the need for changes to reserve operations.

Finally, a small number of patterns related to coping were evident. Spouses who felt that their family coped poorly with the deployment tended to recommend changes to reserve component operations, and service members who similarly described their family were more inclined to cite a need for improved family support programs and resources. On the other hand, both spouses and service members who believed that their family coped well indicated more frequently that they were already doing fine when asked for suggestions to support their family better.

CHAPTER TEN

Conclusion and Recommendations

In this study, we interviewed military family experts as well as both reserve component spouses and service members to provide insights related to how guard and reserve families experience deployment. This approach permitted us to consider not only spouse and service member perceptions related to their deployment experience, but also what implications these perceptions have for family support and service member retention. Specifically, in the report we emphasized family readiness, the problems and positives that stem from deployment, and family coping, and we assessed whether these issues may influence retention intentions. Additional findings pertain to the resources used by families during deployment, possible reasons for their relatively limited use, and spouses and service members' own suggestions for improved family support. Throughout our analysis, we sought to understand differences between the spouses and service members we interviewed as well as differences based on individual or situational characteristics, including pay grade, reserve component, and family financial situation. Indicators of maturity, relationship strength, and experience with military life and deployments were also of note.

One of the strengths of this work is the rich depiction of the types of problems faced by guard and reserve families during deployment as well as the positive aspects of activation and deployment. The analysis of the characteristics that help explain which families experience particular types of problems or positives should guide policymakers as they endeavor to understand and respond to the experiences of reserve component families. In short, we found that the majority of families mention a problem they faced as a result of deployment. However, the kinds of problems and the types of families associated with different problems both differ. For example, while younger spouses and spouses in relatively new marriages were more likely to report emotional or mental problems, spouses with more established families tended to encounter problems with household responsibilities and to report children's issues. The majority of families also mentioned a positive aspect to deployment, such as financial gain, acquired family closeness, patriotism, or increased independence of the spouse or family, and, as with problems, the characteristics of the families likely to report different positives varied.

It is important that policymakers and those organizations chartered to support military families understand the problems encountered and the positives enjoyed by military families, for several reasons. First, DoD has committed to ensuring and promoting general family well-being as part of the Social Compact that recognizes the tremendous sacrifice of military families (MCFP, 2002). Second, not only is family readiness viewed as critical to mission readiness, as discussed earlier in the report, but quality-of-life issues in general are regarded by DoD as inseparable from overall combat readiness (Myers, 2004). Finally, our analysis indicates a relationship between families' problems and positives and military outcomes, including readiness and retention, that affect DoD's ability to satisfy the military mission.

While many of the problems and the positives are compelling, and thus merit short-term attention and the allocation of support resources, the results of our research suggest that successful family support should also be gauged, or even primarily gauged, in terms of family readiness, family coping, and retention intentions—measures of military manpower and family-related outcomes that can guide long-term management of reserve component personnel. While family readiness is widely viewed as important, it has been neither consistently defined nor measured. Similarly, attention has been given to guard and reserve families' ability to cope with activation and deployment, but terminology such as "coping" remains unspecified. Accordingly, in this research effort we asked spouses and service members to explain what readiness and coping meant to their family, and found that both of them were multifaceted constructs. This is of note not only because these concepts are not measured as such, but also because spouses and service members tended to vary in how they defined them.

Although a complicated construct, family readiness emerged as a key factor in our research. We documented a relationship between the individual service member's military preparedness and family readiness, and also found that family readiness was related to most of the deployment-related problems, to a greater likelihood that a family would perceive "no positives" to deployment, to family coping, and to all three measures of retention intention—the impact of the most recent deployment on the service member's military career plans, his or her plans to stay until retirement eligible, and spouse opinion regarding the service member's career intentions. Coping was an important construct as well; it appeared to be related to perceptions of many of the problems and positives, and it was associated with two of the three indicators of retention intention. Yet, the large proportions of spouses and service members who did not provide a definition of coping indicate that it remains an ambiguous concept for families.

Unlike the problems and positives that families reported, there were common patterns across these three interrelated indicators in terms of who tended to respond in ways with favorable implications for family well-being and military effectiveness: being ready or very ready for deployment, coping well or very well, and having retention intentions that reflect a preference for staying. In general, where there are significant patterns, the data suggest that more mature interviewees, those in stronger relationships, and those with prior military experience are more likely to be ready for deployment, to cope well with the deployment, and to indicate a preference for staying in the Guard or Reserve. In addition, those with comfortable family finances and adequate notice of deployment also tended to describe their family as coping well and to report favorable retention intentions.

It is important to note that our exploratory analysis, based on cross-sectional data, did not permit us to address causality, and we did not control for interactions between these indicators. Thus, we have not stated, for example, whether family readiness has a direct effect on the problems or positives experienced, or whether age, pay grade, and marriage length—three potentially interrelated attributes—each have a separate influence on them. Further, although the statistically significant findings reported herein may be applicable to reserve component personnel and spouses with characteristics similar to those who participated in our interviews, additional research is warranted to determine the extent to which these findings are generalizable to *all* families in the Army National Guard, Army Reserve, Marine Forces Reserve, and Air Force Reserve.

All in all, our research indicates that while efforts should be undertaken to avoid or mitigate problems, the family problems stemming from guard and reserve deployments are nuanced, and the solutions are not simple. We found that different families, especially those in different life stages, had varying problems. Different problems have distinct implications and are experienced by different families with varying degrees of severity. This is reinforced by our finding that different families also turned to different kinds of support programs and resources, and that no programs were used by a predominant share of guard and reserve families. Thus, while most of these problems are associated with the important military outcomes of family readiness, family coping, and especially retention intentions, it may be difficult to quantify the effectiveness of programs designed to avoid or mitigate deployment-related problems with respect to these broader outcomes.

Recommendations

Based on the perceptions, experiences, and suggestions for improvement offered by reserve component service members and spouses during interviews, we drafted a set of recommendations. These suggestions are intended to address problems or issues raised by a notable proportion of service members and spouses and that could feasibly be implemented with policy changes. These recommendations are informed by spouse and service member suggestions for improvement, but they neither adopt all those suggestions nor are limited to the interviewees' comments. We view them as constructive steps in the right direction, but we cannot estimate the results of these changes or their cost-effectiveness without further analysis. In some instances, DoD policymakers, including those within OSD and the services, have begun to implement policies and programs consistent with these recommendations. Our research suggests that such actions may prove effective, and our recommendations underscore their importance.

We offer recommendations both to improve family support and also to inform future research as the policies consistent with an operational reserve are fully implemented. The recommendations to improve family support are divided into those related to activation and deployment personnel practices, families' expectations and perceptions, support of and information for families, and measurement of important constructs and outcomes.

Activation and Deployment Personnel Practices

Pursue predictable mobilization in terms of both the length of deployment and the amount of notice. Should the guidelines for one-year involuntary deployments be enacted, it will be easier to ensure predictable mobilization, which is important to service members and their families.

Ensure that any notice sufficient for service members and families to prepare for deployment is also sufficient for the military to receive the entire family. If service members are expected to report to active duty with less than one month's notice, or even less than one week's notice, as was the experience for some of our interview participants, this should also be ample time for the Reserve Component to complete the administrative processes necessary to transition the service member into active duty service and the family members into military programs as appropriate. There should not, for example, be delays in receiving pay for guardsmen and reservists.

Limit the average length of mobilization. Our research suggests that spouses and service members experiencing longer deployments, particularly those one year or longer, were more likely to cope poorly with deployment and to express a preference for leaving the military. Further, one of the suggestions made by interviewed spouses and service members was to reduce deployment length. These findings are consistent with prior research (e.g., DMDC, 2005; Hosek, Kavanagh, and Miller, 2006) that indicated that longer activations were a major source of dissatisfaction for service members and their families. This recommendation is consistent with and emphasizes the significance of announced intentions to limit guard and reserve mobilizations to one year (DoD, 2007). Should this decision be reversed, policymakers should consider whether potential improvements in family-related outcomes—including retention—would offset any possible operational challenges posed by limiting deployment length. The Air Force Reserve and the Marine Forces Reserve, which have average deployments under one year, may offer insights regarding this tradeoff.

In a similar vein, we recommend that DoD **reduce the use of cross-leveling for reserve component personnel.** Our findings suggest that deploying reservists and guardsman separate from the unit that they traditionally drill with may have negative implications for family support, and the detrimental effects that cross-leveling may have for the family were also noted by the CNGR in its Second Report to Congress (2007). Accordingly, DoD's plans to limit this personnel practice (Hall, 2007) should have favorable implications for guard and reserve families.

Perceptions and Expectations

Ensure that family expectations are consistent with the DoD vision of a Reserve Component that is both operational and strategic. Service members and families should recognize that they are likely to begin a new deployment every six years, and that some service members may be tapped to serve more frequently. This includes guardsmen and their families, some of whom had, according to our interviews, not previously realized that their guardsman might be deployed OCONUS for national security reasons.

Recognize that family perceptions are sometimes more important than actual experiences. We found this to be the case with amount of activation notice, where the perceived adequacy of the amount of notice appeared to be a more compelling influence than the actual amount of notice the family received. Further, interviewees had different perceptions of notice adequacy even for the same amount of notice. This suggests that family perceptions of problems and positives, one of the main topics covered in this monograph, may have a greater impact on families than what objective, external measures may indicate.

Recognize that families focus on "boots away from home" and not "boots on the ground." While many manpower experts and policymakers have traditionally focused on "boots on the ground" as the metric for deployment length, guard and reserve families are affected by the entire time the service member is away from home, including pre-deployment and post-deployment training and activities.

Emphasize the positives of activation and deployment. Consistent with prior research (Loughran, Klerman, and Martin, 2006), many of our interviewees experienced an increase in income during their deployment, and some of these financial gains were either unanticipated by the reservist, or they felt that they were unusual in enjoying financial gain. We acknowledge that an increase in income must also cover additional family expenses incurred during the deployment. Nonetheless, DoD could emphasize the financial gain or other positive aspects of deployment in efforts to ensure that reserve component family perceptions of deployment are accurate.

Support of and Information for Families

Increase levels of readiness among not-yet-activated families. Given the likelihood that reserve component service members will be activated at some point in their military career, units should ensure that wills and powers of attorney are regularly updated. Such administrative tasks need not, and should not, wait until a service member is activated.

Know how to find families. DoD should improve the centralized data about families to ensure both notice and information are received in a timely manner. As our expert interviews indicated, this type of information about the location and demographics of guard and reserve families is also critically important to designing and managing appropriate support facilities for families.

Seek ways to provide deployment-phased and "on-demand" information available to families. Given that families continue to ask for more and better information, but also mention the pre-deployment deluge of information, **it is important to tailor both the content and amount of material provided to their needs**. Pre-deployment briefings might be sufficient for some spouses, while they might appear to be a "firehose" of information for spouses unfamiliar with deployments. Focused and intensive workshops might be helpful to some spouses, while others may feel that information from centralized Web sites is sufficient.

Explore ways to connect families to one another, including families that live near one another but represent different units or reserve components. Spouses mentioned the importance of other military spouses during their service member's deployment. Spouses need not be from the same unit, or even the same reserve component, to share the same informal network.

Bear in mind the limited capacity and capabilities of volunteer-based resources, either military or nonmilitary. Many family support organizations, such as Family Readiness Groups, and local community support, such as VFW organizations, depend heavily on volunteers. While the contribution of these individuals is very important, such individuals are not likely to be trained for all possible circumstances and family needs, and such organizations may not be able to continue at the current high level of support for an interminable period of time. DoD should recognize both the strengths and the limitations of these organizations and plan accordingly. This may require DoD, for example, to develop a registry of these organizations, possibly with the assistance of military family advocacy organizations, and to evaluate their resources, strengths, and limitations. Ultimately, metrics pertaining to usage rate, the fraction of the reserve component population served, the fraction seeking assistance but turned away or deterred, and organizational effectiveness in providing support could enable DoD to optimize the "web" of informal, nonmilitary and formal, military resources available to reserve component families.

Consistent with this, and given the reliance that our families reported on nonmilitary resources, **seek ways to improve awareness of, and support or partner with, local and community resources for families**. Creating a family-friendly version of the aforementioned organization registry would be one way to promote this set of resources in a proactive and standardized way, while support and partnerships with local and community resources could take many forms. DoD's relationships with these entities could range from informal to formal or even contractual, and from an endorsement by the DoD to a joint-venture type support effort. DoD's goals, its own resources, and attributes of the local or community resource should guide the determination of the appropriate type of relationship.

Tailor efforts to avoid and mitigate deployment-related problems. Target some support to younger, less experienced military families, but acknowledge that more-established families have other challenges. Recognizing that different kinds of families confront different issues during deployment should help DoD tailor family support resources, especially for units that have disproportionately young families or more established families.

Recognize that, just as the problems experienced by families vary, so do the severity and consequences of problems. Our research was not designed either to discern the varying severity of problems, such as emotional problems or household responsibility issues, or to link problem severity to retention intentions. Nonetheless, family support programs and professionals should be prepared to recognize and handle (or refer) families that suffer more severe problems. Additional research is needed, however, to determine the proportion of families enduring more severe problems and then to allocate resources accordingly.

Consider not only how to help those families that are struggling, but also how to **reinforce and learn from those families who appear to proceed through the deployment cycle with fewer problems.** While all the families included in this research clearly made a significant contribution to the success of the military mission and the United States more broadly, some of the families interviewed appeared to have fewer difficulties overcoming the challenges and problems they faced. Some of these were spouses and service members whose families quietly and successfully endured extended deployments, and were disinclined to focus on the negative aspects of their experiences. DoD should consider ways to learn from these families, such that they might serve as a model or example for other families. For example, DoD might facilitate spouse-mentoring programs, to the extent that spouses who have successfully weathered deployments are willing to share their experiences.

Measurement of Key Constructs and Outcomes

Recognize that family readiness and family coping are multifaceted constructs and develop measures accordingly. We found that current measures of family readiness and family coping do not take into consideration the multiple aspects of these constructs. Given the importance of family readiness and family coping to outcomes such as retention intentions, metrics should be developed that take into account their key dimensions. This would include, for example, their emotional or mental aspects, which more recently married spouses tended to emphasize when asked to define each concept, as well as the household responsibility issues that spouses in more established marriages focused on.

Recognize that service members and spouses may provide different assessments of the same deployment experience. Although we did not interview spouses and service members from the same household, the spouses and service members that we interviewed differed overall in how they view family readiness and coping. This implies that obtaining evaluations of readiness and coping from the spouse only, as has been the typical practice, may provide an incomplete and possibly inaccurate view of a family's condition. We also found differences in spouse and service member tendencies to report various problems, although we could not verify whether their families actually experienced different problems. Thus, our results suggest that relying on either the spouse or the service member to report problems may be insufficient. Obtaining information systematically from both members of a particular household likely would provide a more comprehensive, accurate view of a family's deployment experience, as well as confirm or refute the continued need to do so. Should the need be confirmed, we still acknowledge that this practice may be costly or otherwise difficult to implement on a regular basis. Hence, efforts to better understand, in general, spouse and service member over- and underreporting tendencies, as well as any other "biases," may be adequate to refine the measurement of important constructs such as family readiness.

Use metrics to consider both the short-term and long-term effectiveness of family support. While it is prudent to note success in avoiding and mitigating problems and to track usage rates for particular military programs and services, policymakers should also consider how changes in policies, programs, and services may come to bear on family readiness, family coping, and retention intentions. As discussed earlier, measuring these military manpower and family-related outcomes can be of value for long-term, effective management of reserve component personnel.

Topics for Future Research to Improve the Support for Reserve Component Families

As DoD and other policymakers move forward in supporting reserve component families in an operational reserve, additional research is warranted not only to understand how representative the findings discussed herein are, but also to answer important questions suggested by our research. Topics appropriate to future research include the following:

Explore communities' capacity to support guard and reserve families. In our research we noted references to support provided by the local community, specifically religious organizations, nonprofit organizations, and local business and grassroots efforts. It is not clear, however, whether such organizations—many of which rely on volunteers— have the resources and the energy to sustain high levels of support for extended and repeated deployments. Understanding the range of local support available to guard and reserve families and the capacity of local communities to continue supporting them is important to satisfying the long-term needs of these families. As noted above, some development of meaningful metrics to assess the current and projected capacity of these organizations as well as reserve component families' use of them would be a valuable contribution to DoD's understanding of family support.

Better understand the effect of family readiness on coping and retention. This study noted that families who felt they were ready for the deployment were more likely to report coping well and positive retention plans. The cross-sectional nature of this study prevented us from determining whether readiness had a direct effect on coping or retention plans, and thus this topic deserves further investigation. Future efforts should also seek to understand whether family readiness and coping come to bear not only on career plans but also on actual separations from the Reserve Component.

Explore both the effectiveness and efficiency of solutions intended to improve families' ability to cope during repeat deployments and to help them emerge from deployments with few negative consequences. Because repeat deployments are a feature of today's operationally oriented Reserve Component, understanding how families' needs evolve through repeat deployments and how best to ensure that families develop and maintain good coping skills is critical. This line of research may ultimately include interventions or pilot tests of solutions that are conducted in collaboration with DoD and other family support providers.

Expert Interviews

This research included 26 semi-structured interviews with experts regarding the issues faced by guard and reserve families. Most of the interviews were with single individuals, although others included multiple representatives of the same organization. In all, 15 of the interviews were with experts employed by DoD, and 11 were with representatives of non-DoD groups who advocate for or offer services to military families. The interviews included DoD representatives from OSD and from each of the services who focus on reserve component issues, as well as representatives from the National Guard Bureau. In addition, individuals representing the following organizations were included in the interviews:

- Association of the United States Army (AUSA)
- Enlisted Association of the National Guard of the United States (EANGUS)
- Marine Corps Reserve Association (MCRA)
- Military Officers Association of America (MOAA)
- National Guard Association of the United States (NGAUS)
- National Military Family Association (NMFA)
- Naval Enlisted Reserve Association (NERA)
- Naval Reserve Association (NRA)
- Reserve Enlisted Association (REA)
- Reserve Officers Association (ROA)
- Veterans of Foreign Wars (VFW)

The interviews focused on the problems and positives faced by guard and reserve families during deployment, the resources available to them, and the challenges in delivering those resources. The protocol used for these interviews follows. Note this data collection effort was reviewed and approved by RAND's Institutional Review Board, the entity responsible for ensuring that RAND research adheres to human subjects protection guidelines, such as 45 Code of Federal Regulations (CFR) Part 46 and its subparts (also known as "the Common Rule").

Expert Interview Protocol

1. What is your organization's mission?
2. Broadly speaking, how does your organization support reserve families?
3. What is your position title?

Now that I have some background information about you, let's move into some questions about reserve families. Specifically. . .

4. What types of issues or problems do reserve families have?

   – If subject mentions "needs" (e.g., need for information, need for emotional support), steer toward "problems" by asking, <u>Why do families need XXX? What problem or issue does it resolve?</u>
   – <u>How have these issues/problems changed in recent years?</u>

   Goal is to cover perceptions of reserve family problems (things that happen as a result of deployment) more than what it takes to help reserve families. Consider taking notes in a matrix to match up, where possible, problems with needs, if they kept mentioning needs.

5. How do these issues or problems differ through the deployment cycle, or in other words, before the deployment, during the deployment, and when they return home?

   Probe to determine whether issues mentioned for #5 differ through the deployment cycle (e.g., if she names 3 things in Q5 but only offers difference for 1 in Q6, then we say something like, You also mentioned X. Does this differ through the deployment cycle in any way?)

6. What issues or problems emerge simply due to activation, even without deployment?

7. We've been focusing on the issues and problems that reserve families face. Are there positive outcomes of activation or deployment for reserve families?

   Probe for the descriptions, if they answer "yes."

8. How do reserve families' issues differ from those of active component families?

   We're interested in how either needs or problems differ.

9. What about differences *between* reserve families with different demographics? For example, do younger families have different problems or needs? How so?

   Probe:
   o What about families of female reservists versus those of male reservists?

[Keep in mind the differences between problems and needs and probe as necessary to obtain both, where possible.]

10. Are there differences in family problems or needs depending upon the type of reservist the service member is? For example, do members of different reserve components, such as [relevant

RC example] versus Air National Guard, have different family concerns? *Or* For example, you deal primarily with the [RC]; do members of [different RC], have different family concerns?

[Keep in mind the differences between problems and needs and probe as necessary to obtain both, where possible.]

Probe:
o What about individual augmentees—do they have different family issues than TPU reservists?

11. How does geographic location affect family issues? For example, do reserve families located close to military installations have different issues than those distant from military installations? What about those in rural areas, compared with those in suburban or urban areas?

[Keep in mind the differences between problems and needs and probe as necessary to obtain both, where possible.]

Now let's return to a discussion of support available to guard and reserve families, in particular the resources you mentioned earlier that your organization/[org name] provides.

Prompt, if needed, for an overview as opposed to item-by-item description.

12. Which of your organization's resources are intended especially for reserve families?

13. How important is it to have resources intended specifically for reserve families? Why?

14. *[Ask only if answer to #10 was affirmative.]* You said before that service members from different components have different issues. Are there needs for resources geared to the different components—does the Marine Reserve have different needs than the Air National Guard, for example?

15. What other resources are available to Reserve families for support? In other words, how else do Reserve families resolve the issues and problems we discussed earlier?

    Prompt, if needed, for informal resources: <u>Are there unofficial or informal means of support for reserve families? If yes, please describe</u>.

    Prompt, if not already discussed: <u>What role do local organizations play for Reserve families?</u>

You've provided me with a great deal of useful information today, and we are almost done—we have just a few more questions.

16. What challenges does your organization/[office name] component face in providing reserve families with the support they need?

17. How do you determine whether family support efforts have been successful? *Or* How do you know whether your organization's efforts are successful?

    Probes:
    o <u>*Do you consider whether a family's needs have been met?*</u> *[and how measured?]*

    o <u>*Do you consider whether the negative effects of activation or deployment are avoided or reduced?*</u>

    o <u>*Do you consider how the service member him or herself is affected? For instance, is he deployment ready? Does he intend to stay in the Guard or Reserve?*</u>

    o <u>*What measure—or measures—do you think are the most important? Why?*</u> [alternate—most useful]

18. What opportunities for improvement do you see? How can the DoD better support reserve families? Based on what measures of success?

19. What do you think would be reserve families' biggest complaint about the support they're provided by the military?

20. In closing, is there anything you'd like to add about supporting reserve families? Anything I didn't ask about, but should have?

Consider asking local personnel as interview closes: How do you recommend recruiting spouses for focus groups?

If at any point the interview subject mentions family readiness or preparedness ask, "What is family readiness?" and "How do you assess family readiness?" (If needed, "In other words, how is a family with a high level of readiness different from one with a low level of readiness?")

If subject mentions coping ask, "What exactly is coping?" and "How do you assess coping?" (If needed, "In other words, how is a spouse with a strong ability to cope different from one who has problems coping?")

If subject mentions repeat activations, explore this in greater detail. (e.g., how needs differ in face of repeat activations, how effects differ after repeat activation, whether family resilience/readiness continues to improve with each activation. . .)

Service Member and Spouse Interviews

This appendix features methodological details regarding the interviews conducted with spouses and service members. This information includes data regarding the difficulty contacting our interview sample, the introductory letter that was sent to potential interviewees, the interview introduction, the interview protocols themselves, a technical description of how the interviews were coded and further analyzed, and an abbreviated form of the coding tree used to analyze these interviews.

Difficulty Reaching Potential Interviewees

Landline and Mobile Telephone Numbers

As we discussed in Chapter One, we had difficulty contacting sufficient service members and spouses from the included components, for a couple of reasons. One of the problems we faced was the large proportion of individuals who provided a mobile telephone number in their contact information. Tables B.1 and B.2 include the numbers of telephone numbers we obtained for service members and spouses, and the proportion of those numbers that were mobile telephone numbers. As the tables indicate, this was especially a problem for the junior enlisted personnel and spouses of Air Force Reserve and Marine Forces Reserve, for whom 42 to 44 percent of the telephone numbers provided were cell phone numbers.

Table B.1
Landline and Cell Phone Numbers for Service Members

Pay Grade Category	Army Reserve			Army National Guard		
	Phone	Cell	% Cell	Phone	Cell	% Cell
E1–E4	139	31	22	383	75	20
E5–E6	499	89	18	586	81	14
O1–O3	150	27	18	90	15	17
Totals	788	147	19	1,059	171	16

	Air Force Reserve			Marine Forces Reserve		
	Phone	Cell	% Cell	Phone	Cell	% Cell
E1–E4	72	30	42	141	62	44
E5–E6	610	124	20	120	39	32
O1–O3	38	5	13	44	9	20
Totals	720	159	22	305	110	36

Table B.2
Landline and Cell Phone Numbers for Spouses

Pay Grade Category	Army Reserve			Army National Guard		
	Phone	Cell	% Cell	Phone	Cell	% Cell
E1–E4	238	67	28	535	113	21
E5–E6	607	115	19	852	113	13
O1–O3	124	18	14	183	33	18
Totals	969	200	21	1,570	259	16

	Air Force Reserve			Marine Forces Reserve		
	Phone	Cell	% Cell	Phone	Cell	% Cell
E1–E4	128	55	43	172	76	44
E5–E6	913	219	24	165	46	28
O1–O3	93	20	21	73	24	33
Totals	1,134	294	26	410	146	36

Incorrect Addresses

Another problem that we confronted was incorrect or out-of-date addresses for potential interviewees. Also discussed in Chapter One, Tables B.3 and B.4 provide the number and proportion of letters that were returned due to incorrect addresses.

Table B.3
Introduction Letters Sent to Service Members and Returned

Pay Grade Category	Army Reserve			Army National Guard		
	Sent	Returned	% Returned	Sent	Returned	% Returned
E1–E4	107	32	30	344	39	11
E5–E6	423	76	18	537	49	9
O1–O3	141	9	6	82	8	10
Totals	671	117	5	963	96	10

	Air Force Reserve			Marine Forces Reserve		
	Sent	Returned	% Returned	Sent	Returned	% Returned
E1–E4	203	18	9	128	13	10
E5–E6	589	21	4	115	5	4
O1–O3	71	2	3	43	1	2
Totals	861	41	5	281	19	7

Table B.4
Introduction Letters Sent to Spouses and Returned

Pay Grade Category	Army Reserve			Army National Guard		
	Sent	Returned	% Returned	Sent	Returned	% Returned
E1–E4	224	14	6	493	42	8
E5–E6	577	30	5	822	30	4
O1–O3	123	1	1	179	4	2
Totals	924	45	5	1,494	76	5

	Air Force Reserve			Marine Forces Reserve		
	Sent	Returned	% Returned	Sent	Returned	% Returned
E1–E4	120	8	7	157	15	9
E5–E6	871	42	5	154	11	7
O1–O3	89	4	4	71	2	3
Totals	1,080	54	5	382	28	7

Introduction Letters, Interview Introduction, and Interview Protocol

This section includes the introductory letter sent via U.S. mail to all service members and spouses identified as potential interviewees by their reserve component. The letter was personalized with the individual's name and reserve component, and it differed slightly for service members and for spouses. The second type of item included here is the interview introduction read to each interview participant when he or she was telephoned, including those individuals who subsequently declined to participate. This introduction also differed slightly for service members and spouses. The third type of item in this section is the interview protocols. The interview protocol is a dynamic set of questions, and it directs the interviewer on whether or not to provide the multiple-choice answers and when to skip to other questions, depending on the answers provided. Thus, although the spouse protocol was used for all interviewed spouses, and the service member protocol was used for all interviewed service members, the combination of questions that each participant answered varied because they were based on each individual's situation and answers to prior questions.

Similar to the expert interviews covered in Appendix A, materials included in this section were reviewed and approved by RAND's Institutional Review Board. The project also was subject to a review and licensing process specific to the DoD, and a Report Control Symbol (RCS) was issued, which denotes the interview was part of an official, DoD-approved data collection effort. The RCS number was included in the introductory letter sent to all service members and spouses.

Introduction Letter to Service Members

(RCS) DD-RA(OT)2246
<<date>>

Dear <<service member name>>:

The purpose of this letter is to inform you of a study about Reserve and Guard families being conducted by the RAND Corporation. RAND is a nonprofit public policy research institute. As part of our work for the Department of Defense (DoD), we have been asked by the offices of Reserve Affairs and of Military Community and Family Policy to investigate the perceptions and experiences of Reserve and Guard families regarding activation and deployment.

As part of this study, we are interviewing three types of individuals: 1) spouses of activated Reserve and Guard personnel, 2) spouses of demobilized Reserve and Guard personnel, and 3) demobilized Reserve and Guard personnel with dependents. We have selected a number of Army National Guard, Army Reserve, Air Force Reserve, and Marine Corps Reserve units to participate in this study, including the unit in which you are currently assigned. Your commander is aware of this study and has approved our efforts to conduct interviews. This letter of introduction has been mailed to hundreds of spouses and service members within the [select one: Army National Guard/Army Reserve/Air Force Reserve/Marine Corps Reserve], and a subset of those spouses and service members will be randomly selected to participate in an interview.

If you are selected to participate in an interview, you will be contacted via telephone. During the interview you will be asked questions about your experience as a Reserve service member, your experience with activation and/or deployment(s), and your opinions about both. Taking part in this interview is <u>voluntary</u> and <u>confidential</u>. The commanders of your unit do not know whom we are contacting, nor will they know if you decline to participate. RAND will use the information you provide for research purposes only, and will not disclose your identity or information that identifies you to anyone outside of the project team.

Additionally, because we are interviewing individuals from multiple units, at multiple locations, comments will not be associated with any unit. During the course of the study, we will safeguard the information you provide, and one year after the study is complete we will destroy all information that directly identifies you.

If you have any questions about the study, please contact one of us at the following addresses:

Dr. Laura Castaneda
Management Scientist
RAND Corporation
1776 Main Street
P.O. Box 2138
Santa Monica, CA 90407-2138
Telephone: 310-393-0411,
ext. 6897
Email: Laurawc@rand.org

Dr. Meg Harrell
Senior Social Scientist
RAND Corporation
1200 South Hayes Street
Arlington, VA 22202-5050
Telephone: 703-413-1100,
ext. 5240
Email: Megc@rand.org

You may also contact one of us if you are interested in being interviewed but have not received a telephone call about the study by August 14, 2006. If you have any questions or concerns about your rights as a research subject, you may also contact the Human Subjects Protection Committee at RAND, 1776 Main Street, P.O. Box 2138, Santa Monica, CA 90407-2138, 310-393-0411, ext. 6369.

Thank you for your time and attention. Both DoD and RAND appreciate your support of this important project. The results of this study will be published in a report approximately one year from now. That report will be available from the RAND website at www.rand.org or by request from either of us.

Sincerely,

Dr. Laura Castaneda
Co-Principal Investigator

Dr. Meg Harrell
Co-Principal Investigator

Introduction Letter to Spouses

(RCS) DD-RA(OT)2246
<<date>>

Dear <<spouse name>>:

The purpose of this letter is to inform you of a study about Reserve and Guard families being conducted by the RAND Corporation. RAND is a nonprofit public policy research institute. As part of our work for the Department of Defense (DoD), we have been asked by the offices of Reserve Affairs and of Military Community and Family Policy to investigate the perceptions and experiences of Reserve and Guard families regarding activation and deployment.

As part of this study, we are interviewing three types of individuals: 1) spouses of activated Reserve and Guard personnel, 2) spouses of demobilized Reserve and Guard personnel, and 3) demobilized Reserve and Guard personnel with dependents. We have selected a number of Army National Guard, Army Reserve, Air Force Reserve, and Marine Corps Reserve units to participate in this study, including the unit in which your spouse is currently assigned. Your spouse's commander is aware of this study and has approved our efforts to conduct interviews. This letter of introduction has been mailed to hundreds of spouses and service members within the [select one: Army National Guard/Army Reserve/Air Force Reserve/Marine Corps Reserve], and a subset of those spouses and service members will be randomly selected to participate in an interview.

If you are selected to participate in an interview, you will be contacted via telephone. During the interview you will be asked questions about your experience as a Reserve spouse, your experience with activation and/or deployment(s), and your opinions about both. Taking part in this interview is <u>voluntary</u> and <u>confidential</u>. The commanders of your spouse's unit do not know whom we are contacting, nor will they know if you decline to participate. RAND will use the information you provide for research purposes only, and will not disclose your identity or information that identifies you to anyone outside of the project team.

Additionally, because we are interviewing individuals from multiple units, at multiple locations, comments will not be associated with any unit. During the course of the study, we will safeguard the information you provide, and one year after the study is complete we will destroy all information that directly identifies you.

If you have any questions about the study, please contact one of us at the following addresses:

Dr. Laura Castaneda
Management Scientist
RAND Corporation
1776 Main Street
P.O. Box 2138
Santa Monica, CA 90407-2138
Telephone: 310-393-0411,
ext. 6897
Email: Laurawc@rand.org

Dr. Meg Harrell
Senior Social Scientist
RAND Corporation
1200 South Hayes Street
Arlington, VA 22202-5050
Telephone: 703-413-1100,
ext. 5240
Email: Megc@rand.org

You may also contact one of us if you are interested in being interviewed but have not received a telephone call about the study by August 14, 2006. If you have any questions or concerns about your rights as a research subject, you may also contact the Human Subjects Protection Committee at RAND, 1776 Main Street, P.O. Box 2138, Santa Monica, CA 90407-2138, 310-393-0411, ext. 6369.

Thank you for your time and attention. Both DoD and RAND appreciate your support of this important project. The results of this study will be published in a report approximately one year from now. That report will be available from the RAND website at www.rand.org or by request from either of us.

Sincerely,

Dr. Laura Castaneda
Co-Principal Investigator

Dr. Meg Harrell
Co-Principal Investigator

Interview Introduction to Service Members

Introduction:

Good morning/afternoon/evening. I'm calling on behalf of the RAND Corporation study of guard and reserve families.

May I please speak to [SERVICE MEMBER NAME]? [IF INITIAL CALLED PARTY IS NOT SERVICE MEMBER, REPEAT]

Good morning/afternoon/evening. I'm calling on behalf of the RAND Corporation study of guard and reserve families.

You may remember from a letter recently mailed to you that RAND is a nonprofit organization that conducts research for the Department of Defense. As part of this research, RAND has been asked to investigate the perceptions and experiences of guard and reserve families regarding activation and deployment.

You may also recall from RAND's letter of introduction that your unit is one of the units approved for study within the [select one: Army National Guard/Army Reserve/Air Force Reserve/Marine Corps Reserve].

My name is _____. I am from SRBI, a company RAND has employed to conduct interviews for this study. We are now in the process of contacting a group of randomly selected service members from that unit, and that's the reason for my call today. If you agree to be interviewed, either today or at a more convenient time, I will be asking you questions about your experience with activation and/or deployment and your opinions about both.

Taking part in this interview is voluntary. Please let me know if you don't want to participate in this interview, or if you want to stop it at any time and for any reason. You should also feel free to skip any questions that you prefer not to answer.

In addition, this interview is <u>confidential</u>. The commanders of your unit are aware of our research, but do not know whom we are contacting, nor will they know if you decline to participate.

I will take notes during our conversation, but I will not insert your name into the notes. With your permission, I will also record parts of our conversation so that I accurately capture your responses.

RAND will use the information you give me for research purposes only, and will not disclose your identity or information that identifies you to anyone outside of the project team, except as required by law. Additionally, because we are interviewing individuals from multiple units, at multiple locations, comments will not be associated with any unit. During the course of the study, the project team will safeguard the information you provide, and one year after the study is complete, all information that directly identifies you will be destroyed.

The interview will take approximately 30 minutes.

Do you have any questions about the study?

Do you need a copy of RAND's letter of introduction sent to you again?

Do you want to be interviewed now, would you like to schedule an interview for a more convenient time, or do you want to wait until you receive RAND's letter before deciding?

Do you agree to participate in this research interview?

Interview Introduction to Spouses

Introduction:

Good morning/afternoon/evening. I'm calling on behalf of the RAND Corporation study of guard and reserve families.

May I please speak to [SPOUSE NAME]? [IF INITIAL CALLED PARTY IS NOT SPOUSE, REPEAT]

Good morning/afternoon/evening. I'm calling on behalf of the RAND Corporation study of guard and reserve families.

You may remember from a letter recently mailed to you that RAND is a nonprofit organization that conducts research for the Department of Defense. As part of this research, RAND has been asked to investigate the perceptions and experiences of guard and reserve families regarding activation and deployment.

You may also recall from RAND's letter of introduction that your unit is one of the units approved for study within the [select one: Army National Guard/Army Reserve/Air Force Reserve/Marine Corps Reserve].

My name is _____. I am from SRBI, a company RAND has employed to conduct interviews for this study. We are now in the process of contacting a group of randomly selected service members from that unit, and that's the reason for my call today. If you agree to be interviewed, either today or at a more convenient time, I will be asking you questions about your experience with activation and/or deployment and your opinions about both.

Taking part in this interview is voluntary. Please let me know if you don't want to participate in this interview, or if you want to stop it at any time and for any reason. You should also feel free to skip any questions that you prefer not to answer.

In addition, this interview is <u>confidential</u>. The commanders of your unit are aware of our research, but do not know whom we are contacting, nor will they know if you decline to participate.

I will take notes during our conversation, but I will not insert your name into the notes. With your permission, I will also record parts of our conversation so that I accurately capture your responses.

RAND will use the information you give me for research purposes only, and will not disclose your identity or information that identifies you to anyone outside of the project team, except as required by law. Additionally, because we are interviewing individuals from multiple units, at multiple locations, comments will not be associated with any unit. During the course of the study, the project team will safeguard the information you provide, and one year after the study is complete, all information that directly identifies you will be destroyed.

The interview will take approximately 30 minutes.

Do you have any questions about the study?

Do you need a copy of RAND's letter of introduction sent to you again?

Do you want to be interviewed now, would you like to schedule an interview for a more convenient time, or do you want to wait until you receive RAND's letter before deciding?

Do you agree to participate in this research interview?

Interview Protocol for Service Members

Supporting and Retaining Guard and Reserve Families: Interview Protocol for Demobilized Reserve Component Personnel

Service Member Background
"Let me start today by asking you some background questions."

1. Were you activated for at least one month since September 11, 2001?

   1. Yes.
   2. No => Thank for participation and terminate interview.
   98. Don't know => Thank for participation and terminate interview.
   99. Prefer not to answer => Thank for participation and terminate interview.

2. Did you deploy outside the contiguous 48 states, or OCONUS, at least once since September 11, 2001?

   1. Yes
   2. No => Thank for participation and terminate interview.
   98. Don't know => Thank for participation and terminate interview.
   99. Prefer not to answer => Thank for participation and terminate interview.

3. What is your present pay grade in the [auto-insert specific reserve component]? [Interviewer: DON'T READ CHOICES.]

   1. E-1
   2. E-2
   3. E-3
   4. E-4
   5. E-5
   6. E-6

7. E-7 => Thank for participation and terminate interview.
8. E-8 => Thank for participation and terminate interview.
9. E-9 => Thank for participation and terminate interview.
10. O-1/O-1E
11. O-2/O-2E
12. O-3/O-3E
13. O-4 => Thank for participation and terminate interview.
14. O-5 => Thank for participation and terminate interview.
15. O-6 and above => Thank for participation and terminate interview.
98. Don't know => Thank for participation and terminate interview.
99. Prefer not to answer => Thank for participation and terminate interview.

4. Did you previously serve in an active-duty Service (for example, Army, Navy, Air Force, Marine Corps) for 2 years or more?

1. Yes
2. No
98. Don't know
99. Prefer not to answer

5. How many years have you spent in military service? Include both time spent as an active-duty service member and time spent in a National Guard or Reserve component. [Interviewer: ROUND UP TO COMPLETE YEARS. If subject doesn't recall total years, first suggest that he/she estimate before selecting the don't know option.]

_____ years (Range: 0–50, 98, 99) (Less than 1 year record as "0")

[Interviewer: If asked, the following count toward military time:
• Time spent as an active duty service member
• Time spent drilling as a drilling unit Reservist/Traditional Guardsman
• Time spent mobilized or activated on active duty

- Time spent in a full-time active duty program
- Time spent in Individual Ready Reserves (IRR)
- Time spent as an Individual Mobilization Augmentee (IMA)]

98. Don't know
99. Prefer not to answer

6. During your most recent activation, were you an Individual Mobilization Augmentee?

[Interviewer: If subject is unsure, ask: Were you part of a unit that trained together and then was activated together, (If "yes" then enter punch <2> No), or were you activated to join a unit already in place? (If "yes" then enter punch <1> Yes) (The latter option is an Individual Mobilization Augmentee.)]

1. Yes
2. No
98. Don't know
99. Prefer not to answer

7. [Interviewer: RECORD GENDER BY OBSERVATION. If necessary, ask:] What is your gender? [Interviewer: DON'T READ CHOICES.]

1. Male
2. Female
99. Prefer not to answer

8. How old were you on your last birthday?

_____ years (Range 18–70, 99) [CATI: ALLOW < 18 TO BE ENTERED, Thank for participation and terminate interview.]

99. Prefer not to answer

[Interviewer: If service member prefers not to answer, "99" => Q8A]

8A. Are you 18 years or older?

1. Yes
2. No => Thank for participation and terminate interview.
99. Prefer not to answer => Thank for participation and terminate interview.

9. Are you Spanish, Hispanic, or Latino/Latina [select correct term based on Q7 response]? [Interviewer: DON'T READ LIST. If needed to clarify, ask: Are you Mexican, Mexican-American, Chicano, Puerto Rican, Cuban, or another type of Spanish, Hispanic, or Latino/Latina origin or descent?]

1. Yes
2. No
98. Don't know
99. Prefer not to answer

10. What is your race? [Interviewer: IF INITIAL ANSWER DOES NOT CLEARLY MAP TO ONE OF OPTIONS, READ LIST. SELECT ALL THAT APPLY]

1. White
2. Black or African American
3. American Indian or Alaska Native
4. Asian (e.g., Asian Indian, Chinese, Filipino, Japanese, Korean, Vietnamese)
5. Native Hawaiian or other Pacific Islander (e.g., Samoan, Guamanian, or Chamorro)
98. Don't know
99. Prefer not to answer

Education and Employment

11. What is the highest degree or level of school that you have completed? [Interviewer: DON'T READ LIST.]

1. 12 years or less of school (no diploma)
2. High school graduate—high school diploma or equivalent (GED)
3. Some college credit, but less than 1 year
4. 1 or more years of college, no degree
5. Associate's degree (includes AA, AS)
6. Bachelor's degree (includes BA, AB, BS)
7. Master's, doctoral, or professional school degree (includes MA, MS, MEng, MBA, MDE, PhD, MD, JD, DVM)
99. Prefer not to answer

12. If you are currently enrolled in school, what kind of school are you enrolled in? [Interviewer: DON'T READ LIST.]

1. Does not apply; not currently enrolled in school
2. High school
3. Vocational school
4. 2-year college
5. Undergraduate program at 4-year college or university
6. Post-bachelor's degree program leading to a master's, doctoral, or professional degree
7. Other: _____
99. Prefer not to answer

13. Which of the following categories best describes your current civilian employment status? You can stop me when you hear the appropriate category. [Interviewer: READ LIST.]

1. Employed full-time (35 or more hours per week)
2. Employed part-time (less than 35 hours per week)
3. Not employed, but seeking full-time or part-time employment => Q17

4. Not employed and not currently looking for employment => Q16
99. Prefer not to answer => Q17

14. Are you self-employed?

1. Yes
2. No
99. Prefer not to answer

15. Are you in a family business?

1. Yes
2. No
99. Prefer not to answer

[SKIP TO Q17]

16. Would you tell me why this is your choice for now? [WRITE ANSWER]

1. Gave Response
98. Don't know
99. Prefer not to answer

17. In what ways, if any, has your service in the National Guard/Reserve affected your civilian employment or education? [RECORD]

1. Gave Response
98. Don't know
99. Prefer not to answer

Family Background
"Thank you. Now I have some questions about your family."

18. How many children or legal dependents under the age of 18 do you have?

_____ children (Range 0–12, 99)

99. Prefer not to answer

If 0 or prefer not to answer => Q21

19. What are their ages? (Range 0–17, 99)

Child 1: _____ years
Child 2: _____ years
Child 3: _____ years
Child 4: _____ years
Child 5: _____ years
Child 6: _____ years
Child 7: _____ years

Add additional fields as needed for 12 or more children.

99. Prefer not to answer

[Interviewer: Enter 0 for child under age 1.]

20. Which of those children reside with you either full-time or part-time when you're not deployed?

[Auto-insert first two columns of this table from Q19 response]

Child 1: _____ years 1. YES 2. NO 99. Prefer not to answer
Child 2: _____ years 1. YES 2. NO 99. Prefer not to answer
Child 3: _____ years 1. YES 2. NO 99. Prefer not to answer
Child 4: _____ years 1. YES 2. NO 99. Prefer not to answer
Child 5: _____ years 1. YES 2. NO 99. Prefer not to answer

Child 6: _____ years 1. YES 2. NO 99. Prefer not to answer
Child 7: _____ years 1. YES 2. NO 99. Prefer not to answer

Add additional fields as needed for up to 12 children.

    99. Prefer not to answer

21. What is your marital status? [Interviewer: DON'T READ LIST. If participant responds with "Single," ask: "Are you separated, divorced, widowed or have you never married?"]

    1. Married => Q24
    2. Separated
    3. Divorced
    4. Widowed
    5. Never married
    99. Prefer not to answer

22. Do you have a significant other or girlfriend/boyfriend?

    1. Yes
    2. No => If Q18= 0 or prefer not to answer, then thank for participation and terminate interview; otherwise SKIP TO Q28.
    99. Prefer not to answer => If Q18= 0 or prefer not to answer, then thank for participation and terminate interview; otherwise SKIP TO Q28.

23. How many years have you been in a relationship with your significant other or girlfriend/boyfriend?

_____ years (Range 0–50, 98, 99) (Less than 1 year record as "0")

    98. Don't know
    99. Prefer not to answer

[SKIP TO Q25]

24. For how many years have you and your spouse been married?

_____ years (Range 0-50, 98, 99) (Less than 1 year record as "0")

   98. Don't know
   99. Prefer not to answer

25. What is the highest degree or level of school that your spouse/significant other has completed? [Interviewer: DON'T READ LIST.]

1. 12 years or less of school (no diploma)
2. High school graduate—high school diploma or equivalent (GED)
3. Some college credit, but less than 1 year
4. 1 or more years of college, no degree
5. Associate's degree (includes AA, AS)
6. Bachelor's degree (includes BA, AB, BS)
7. Master's, doctoral, or professional school degree (includes MA, MS, MEng, MBA, MDE, PhD, MD, JD, DVM)
98. Don't know
99. Prefer not to answer

26. Which of the following categories best describes your spouse's/significant other's current employment status? You can stop me when you hear the appropriate category. [Interviewer: READ LIST.]

1. Employed full-time (35 or more hours per week)
2. Employed part-time (less than 35 hours per week)
3. Not employed, but seeking full-time or part-time employment => Q28
4. Not employed and not currently looking for employment => SKIP TO Q28
99. Prefer not to answer => SKIP TO Q28

27. How much does your spouse's/significant other's employment income contribute toward your total monthly household income? Please choose one of the following four options.

[Interviewer: READ LIST. If needed, define monthly household income as income earned by you and your spouse/significant other from all sources in a typical month.]

1. No contribution
2. Minor contribution
3. Moderate contribution
4. Major contribution
98. Don't know
99. Prefer not to answer

28. Which of the following best describes your family's current financial situation? Please choose one of the following five options. [Interviewer: READ LIST.]

1. Very comfortable and secure
2. Able to make ends meet without much difficulty
3. Occasionally have some difficulty making ends meet
4. Tough to make ends meet but keeping your head above water
5. In over your head
98. Don't know
99. Prefer not to answer

Activation and Deployment
"Now I have some questions about your recent service as a member of the National Guard/Reserve and how it affected your family."

29. Since September 11, 2001, how many times have you been activated, or in other words, called to active duty? We will discuss deployments in a couple of minutes.

_____ times (Range 1–8, 98, 99)

(CATI: 1-8 Skip to Q30)

98. Don't know
99. Prefer not to answer

29A. Since September 11, 2001, were you activated more than once?

1. Yes
2. No => SKIP TO Q30
98. Don't know
99. Prefer not to answer => SKIP TO Q31

[Interviewer: For duration portion of the question, round up partial months—e.g., 1 month, 1 day should be entered in as 2 months. If they provide duration as days, weeks, years, then translate to months. For recency portion, start date (month/years) is the preferred answer format, but answers such as 'about 15 months ago' are acceptable. If reservist doesn't know part of the answer, first suggest that he/she estimates duration and/or recency before selecting the don't know option. E.g., "Could you estimate how long the deployment was" or "Could you estimate how long ago the deployment was?"]

29B.1 How long was the most recent activation?

_____ Months (Range 0–48, 98, 99)

98. Don't know
99. Prefer not to answer

29B.2 In what month and year did the most recent activation begin?

1. Gave Month and Year => Month _____ (1–12) Year _____ (2001–2006) => Skip to Q31
98. Don't know => Ask Q29B.3
99. Prefer not to answer => Ask Q29B.3

29B.3 How long ago did it begin?

_____ Months (Range 0–48, 98, 99)

98. Don't know
99. Prefer not to answer

[SKIP TO Q31]

[CATI: ASK FOR UP TO 8 ITERATIONS/ACTIVATIONS]

[Interviewer: For duration portion of the question, round up partial months—e.g., 1 month, 1 day should be entered in as 2 months. If they provide duration as days, weeks, years, then translate to months. For recency portion, start date (month/years) is the preferred answer format, but answers such as 'about 15 months ago' are acceptable. If reservist doesn't know part of the answer, first suggest that he/she estimates duration and/or recency before selecting the don't know option. E.g., "Could you estimate how long the activation was" or "Could you estimate how long ago the activation was?"]

30.1 [If 1 activation]: How long was the activation?

[If more than 1 activation:] How long was the ### activation?

_____ Months (Range 0–48, 98, 99)

98. Don't know
99. Prefer not to answer

30.2 [If 1 activation: In what month and year did the activation begin?

[If more than 1 activation:] In what month and year did the ### activation begin?

1. Gave Month and Year => Month _____ (1–12) Year _____ (2001–2006) => Skip to Q31
98. Don't know => Ask Q30.3
99. Prefer not to answer => Ask Q30.3

30.3 [If 1 activation:] How long ago did it begin?

[If more than 1 activation:] How long ago did the ### activation begin?

_____ Months (Range 0–48, 98, 99)

  98. Don't know
  99. Prefer not to answer

31. How far in advance did you receive notice of your most recent activation before you reported for active duty?

_____ [Interviewer: Looking for a duration specified in units of time—e.g., hours, days, or weeks]

  1.   Gave answer in Hours => _____ Hours (Range 0–96)
  2.   Gave answer in Days => _____ Days (Range 1–90)
  3.   Gave answer in Weeks => _____ Weeks (Range 1–52)
  4.   Gave answer in Months => _____ Months (Range 1–24)
  98. Don't know
  99. Prefer not to answer

32. Did the amount of notice affect how well prepared your family was for your activation? If yes, how? [RECORD]

  1.   Gave response
  98. Don't know
  99. Prefer not to answer

33. How far does your family currently live from the place where your unit regularly drills or trains? [Interviewer: Obtain estimate in miles; round up as needed.]

_____ miles (Range 0–997, 998, 999)

  998. Don't know
  999. Prefer not to answer

34. How far is it, one way, to the nearest military installation from your family's residence? [Interviewer: Obtain estimate in miles; round up as needed. If asked, military installation may belong to any Service; it does not have to correspond to the service member's Service (e.g., an Army Reserve spouse's closest military installation could be an Air Force base.]

_____ miles (Range 0–997, 998, 999)

   998. Don't know
   999. Prefer not to answer

35. Is your family's current residence more than 10 miles away from where your family resided during your most recent activation? [Interviewer: Enter punch <2> No if respondent did not move or moved 10 or less miles]

   1. Yes
   2. No => SKIP TO Q36
   98. Don't know => SKIP TO Q36
   99. Prefer not to answer => SKIP TO Q36

35A. During your most recent activation, how far did your family live from the place where your unit regularly drills or trains? [Interviewer: Obtain estimate in miles; round up as needed. If needed in response to question or confusion, add: "I am looking for the driving distance your family members would have had to travel to get to your drill location during your most recent activation, even though you may not have been there."]

_____ miles (Range 0–997, 998, 999)

   998. Don't know
   999. Prefer not to answer

35B. During your most recent activation, how far was it, one way, to the nearest military installation from your residence? [Interviewer: Obtain estimate in miles; round up as needed. If needed in response to question or confusion, add: "I am looking for the driving distance your family members would have had to travel, during your most recent activation, to get from their residence to the nearest military installation."]

_____ miles (Range 0–997, 998, 999)

   998. Don't know
   999. Prefer not to answer

36. Since September 11, 2001, how many times have you been deployed outside the contiguous 48 states, or OCONUS?

_____ times => (Range 1–8, 98, 99) (CATI: 1-8 Skip to Q37)

   98. Don't know
   99. Prefer not to answer

36A. Since September 11, 2001, did you deploy outside the contiguous 48 states more than once?

   1. Yes
   2. No => SKIP TO Q37
   98. Don't know
   99. Prefer not to answer => SKIP TO Q38

[Interviewer: For duration portion of the question, round up partial months—e.g., 1 month, 1 day should be entered in as 2 months. If they provide duration as days, weeks, years, then translate to months. For recency portion, start date (month/years) is the preferred answer format, but answers such as 'about 15 months ago' are acceptable. If reservist doesn't know part of the answer, first suggest that he/she estimates duration and/or recency before selecting the don't know option. E.g., "Could you estimate how long the deployment was" or "Could you estimate how long ago the deployment was?"]

36B.1 How long was the most recent deployment?

_____ Months (Range 0–48, 98, 99)

   98. Don't know
   99. Prefer not to answer

36B.2 In what month and year did the most recent deployment begin?

   1.   Gave Month and Year => Month ___ (1–12) Year ___ (2001–2006) => Skip to CONDITION BEFORE Q38
   98. Don't know => Ask Q36B.3
   99. Prefer not to answer => Ask Q36B.3

36B.3 How long ago did it begin?

_____ Months (Range 0–48, 98, 99)

   98. Don't know
   99. Prefer not to answer

[SKIP TO CONDITION BEFORE Q38]

[CATI: ASK FOR UP TO 8 ITERATIONS/DEPLOYMENTS]
[Interviewer: For duration portion of the question, round up partial months—e.g., 1 month, 1 day should be entered in as 2 months. If they provide duration as days, weeks, years, then translate to months. For recency portion, start date (month/years) is the preferred answer format, but answers such as 'about 15 months ago' are acceptable. If reservist doesn't know part of the answer, first suggest that he/she estimates duration and/or recency before selecting the don't know option. E.g., "Could you estimate how long the deployment was" or "Could you estimate how long ago the deployment was?"]

37.1[If 1 deployment]: How long was the deployment?

[If more than 1 deployment:] How long was the ### deployment?

_____ Months (Range 0–48, 98, 99)

  98. Don't know
  99. Prefer not to answer

37.2 [If 1 deployment:] In what month and year did the deployment begin?

[If more than 1 deployment:] In what month and year did the ### deployment begin?

  1. Gave Month and Year => Month ___ (1–12) Year ___ (2001–2006) => Skip to CONDITION BEFORE Q38
  98. Don't know => Ask Q37.3
  99. Prefer not to answer => Ask Q37.3

37.3 [If 1 deployment:] How long ago did it begin?
[If more than 1 deployment:] How long ago did the ### deployment begin?

_____ Months (Range 0–48, 98, 99)

  98. Don't know
  99. Prefer not to answer

[IF Q36 = 2-8 OR Q36A = 1 ASK Q38, ELSE SKIP TO CONDITION BEFORE Q39]

38. Were you employed by the same employer through multiple deployments?

  1. Yes
  2. No
  97. Does not apply; I did not work through multiple deployments.
  98. Don't know
  99. Prefer not to answer

[IF Q28 = 98, 99 SKIP TO Q41]

39. You described your current financial situation as [auto-insert from Q28]. Has this changed because of your activation or deployment?

   1. Yes
   2. No => SKIP TO Q41
  98. Don't know => SKIP TO Q41
  99. Prefer not to answer => SKIP TO Q41

40. Did your financial situation change for the better or for the worse?

   1. For the better
   2. For the worse
  98. Don't know
  99. Prefer not to answer

41. What does it mean for your family to be READY for activation or deployment? [RECORD]

   1. Gave response
  98. Don't know
  99. Prefer not to answer

42. How READY was your family for your most recent deployment? [WRITE ANSWER]

   1. Gave response
  98. Don't know
  99. Prefer not to answer

43. What types of issues or problems did your family face, or is currently facing, as a result of your activation or deployment? [RECORD]

[Interviewer: Always use one of the following two probes after first giving the subject a chance to respond:

1) *Probe if subject expresses confusion or does not offer a response after a short pause:*

Issues that have been mentioned to us include those related to emotional stability, health care, employment for the service member *or* the spouse, education for the service member *or* the spouse, family finances, household responsibilities and chores, marital health, and children.

Can you talk about the extent to which any of these have been issues with your family?

2) *Probe if subject answers the question and first probe was not used:*

Issues that have been mentioned to us include those related to emotional stability, health care, employment for the service member *or* the spouse, education for the service member *or* the spouse, family finances, household responsibilities and chores, marital health, and children.

Is there anything you'd like to add to your response of this question?

1. Gave response
98. Don't know
99. Prefer not to answer

44. What do you think COPING with activation or deployment means for your family? [RECORD]

1. Gave response
98. Don't know
99. Prefer not to answer

45. How well has your family COPED with your most recent deployment, and why do you say that? [WRITE ANSWER] [Interviewer: Ensure both parts of question are answered. Probe if participant has not included both how well his/her family has coped and "why" in answer.]

1. Gave response
98. Don't know
99. Prefer not to answer

46. In what ways, if any, has your most recent activation or deployment been a positive experience for your family? [RECORD]

[Interviewer: Prompt if needed: What positives or good things have come about as a result of your activation or deployment?]

1. Gave response
98. Don't know
99. Prefer not to answer

[If Q36 = 1 OR Q36A = No, Don't know, or Prefer not to state => SKIP TO Q50]

Repeat Deployments

47. How did your family's experience with your first deployment differ from that of your most recent deployment? [RECORD]

1. Gave response
98. Don't know
99. Prefer not to answer

[IF Q21 = 1 OR Q22 = 1 ASK Q48, ELSE SKIP TO Q49]

48. How did your multiple deployments since September 11, 2001, affect your spouse's/significant other's work or education? [RECORD]

1. Gave response
98. Don't know
99. Prefer not to answer

[SKIP TO Q50. Service members with a spouse or significant other should not be asked Q49.]

49. How did your multiple deployments since September 11, 2001, affect your role or your responsibilities as a single parent? [RECORD]

1. Gave response
98. Don't know
99. Prefer not to answer

Support Resources
"My next set of questions pertains to resources that Guard and Reserve families may turn to for support during activation and deployment."

50. Military-sponsored family support programs offer services to National Guard/Reserve personnel and their families, particularly during activation and deployments. Such services include the Family Readiness Group, Military OneSource financial or legal counseling, and assistance with TRICARE. Was your family aware of these military-sponsored programs during your most recent activation?

1. Yes
2. No => SKIP TO Q53
98. Don't know => SKIP TO Q53
99. Prefer not to answer => SKIP TO Q53

51. Did your family participate in or use such a program during your most recent activation?

1. Yes
2. No => SKIP TO Q53
98. Don't know => SKIP TO Q53
99. Prefer not to answer => SKIP TO Q53

52. What types of programs or services did your family use? [WRITE ANSWER]

   1. Gave response
   98. Don't know
   99. Prefer not to answer

53. What nonmilitary, informal, or community resources did your family turn to or use during your most recent activation? [RECORD]

[Interviewer: Always use one of the following two probes after first giving the subject a chance to respond:

1) *If subject expresses confusion or does not offer a response after a short pause:*

Informal or nonmilitary resources that have been mentioned to us include those such as extended family, church, and organizations in your community, like the VFW or the Red Cross.

Can you talk about the extent to which your family used any of these resources?

2) *If subject answers the question and first probe was not used:*

Informal or nonmilitary resources that have been mentioned to us include those such as extended family, church, and organizations in your community, like the VFW or the Red Cross.

Is there anything you'd like to add to your response to this question?

   1. Gave response
   98. Don't know
   99. Prefer not to answer

54. Do you know what resources, either military-sponsored, nonmilitary, and/or informal were the most useful to your family? Which ones

and why? [RECORD] [Interviewer: Ensure both parts of question are answered. Probe if participant has not included both resources and "why" in answer]

1. Gave response
98. Don't know
99. Prefer not to answer

Retention
"Thank you for your thoughtful responses to my questions about your family. We are almost done. My next set of questions pertains to *your* participation in the National Guard/Reserve."

55. Overall, how well prepared were you to perform your active duty job during your most recent activation? Please choose one of the following five options. [Interviewer: READ LIST.]

1. Very well prepared
2. Well prepared
3. Neither well nor poorly prepared
4. Poorly prepared
5. Very poorly prepared
98. Don't know
99. Prefer not to answer

56. At the present time, which statement best describes your National Guard/Reserve career plans? You can stop me when you hear the appropriate category. [Interviewer: READ LIST AS FAR AS NEEDED.]

1. To leave the National Guard/Reserve before completing your present obligation
2. To stay in the National Guard/Reserve and leave after you complete your present obligation
3. To stay in the National Guard/Reserve beyond your present obligation, but not necessarily until you qualify for retirement

4. To stay in the National Guard/Reserve until you qualify for retirement, but not until mandatory retirement age
5. To stay in the National Guard/Reserves until you reach mandatory retirement age
98. Undecided
99. Prefer not to answer

57. What impact has your recent activation had on your National Guard/Reserve career intentions? Please choose one of the following five options. [Interviewer: READ LIST AS FAR AS NEEDED.]

1. Greatly increased your desire to stay
2. Increased your desire to stay
3. Has no influence
4. Increased your desire to leave
5. Greatly increased your desire to leave
98. Don't know
99. Prefer not to answer

[IF Q21 = 1 OR Q22 = 1 ASK Q58, ELSE SKIP TO Q59]

58. Does your spouse/significant other think you should stay on or leave the National Guard/Reserve? Please choose one of the following five options. You can stop me when you hear the appropriate option. [Interviewer: READ LIST AS FAR AS NEEDED]

1. She/he strongly favors your staying
2. She/he somewhat favors your staying
3. She/he has no opinion one way or the other
4. She/he somewhat favors your leaving
5. She/he strongly favors your leaving
98. Don't know
99. Prefer not to answer

Closing Questions

59. In closing, how can the military provide better support for you and your family? [RECORD]

   1. Gave response
  98. Don't know
  99. Prefer not to answer

60. Are there other comments you'd like to make regarding the topics we discussed today?

   1. Yes => Write the comments
   2. No => Thank for participation and end interview
  99. Prefer not to answer

Interview Protocol for Spouses

Supporting and Retaining Guard And Reserve Families:
Interview Protocol for Spouses Of Activated or Demobilized Reserve Component Personnel

Service Member Background

"First I have some questions about your spouse, which will ensure I ask you the right questions today. As we progress through the interview, I have questions about *your* background and your perceptions and opinions."

1. Is your spouse or significant other currently a member of the National Guard or Reserve?

1. Yes
2. No => Thank for participation and terminate interview. [Interviewer: This includes individuals who indicate they are now separated or divorced from a service member.]
98. Don't know => Thank for participation and terminate interview.
99. Prefer not to answer => Thank for participation and terminate interview.

"For simplicity, from this point forward I'm going to refer to your spouse or significant other as your 'spouse.'"

2. Was your spouse activated, or in other words, called to active duty, for at least one month since September 11, 2001?

1. Yes
2. No => Thank for participation and terminate interview.
98. Don't know => Thank for participation and terminate interview.
99. Prefer not to answer => Thank for participation and terminate interview.

3. Did your spouse deploy outside the contiguous 48 states, or OCONUS, at least once since September 11, 2001?

   1. Yes
   2. No => Thank for participation and terminate interview.
  98. Don't know => Thank for participation and terminate interview.
  99. Prefer not to answer => Thank for participation and terminate interview.

4. What is your spouse's present pay grade in the [auto-insert specific Reserve Component]? [Interviewer: DON'T READ CHOICES.]

   1. E-1 => Q5
   2. E-2 => Q5
   3. E-3 => Q5
   4. E-4 => Q5
   5. E-5 => Q5
   6. E-6 => Q5
   7. E-7 => Thank for participation and terminate interview.
   8. E-8 => Thank for participation and terminate interview.
   9. E-9 => Thank for participation and terminate interview.
  10. O-1/O-1E => Q5
  11. O-2/O-2E => Q5
  12. O-3/O-3E => Q5
  13. O-4 => Thank for participation and terminate interview.
  14. O-5 => Thank for participation and terminate interview.
  15. O-6 and above => Thank for participation and terminate interview.
  98. Don't know
  97. Prefer not to answer => Thank for participation and terminate interview.

4A. Our records indicate that his/her present pay grade is [auto insert pay grade]. Does that sound right to you, or can you recall a title like staff sergeant or captain?

    1. Yes, that pay grade is correct.
    2. Other: [WRITE ANSWER]
    98. Don't know
    99. Prefer not to answer

5. Did your spouse previously serve in an active-duty Service (for example, Army, Navy, Air Force, or Marine Corps) for 2 years or more?

    1. Yes
    2. No
    98. Don't know
    99. Prefer not to answer

6. How many years has your spouse spent in military service? Include both time spent as an active-duty service member and time spent in a National Guard or Reserve component. [Interviewer: ROUND UP TO COMPLETE YEARS. If spouse doesn't know total years, first suggest that he/she estimate before selecting the don't know option.]

_____ years (Range: 0–50, 98, 99) (Less than 1 year record as "0")

[Interviewer: If asked, the following count toward military time:
- Time spent as an active duty service member
- Time spent drilling as a drilling unit Reservist/Traditional Guardsman
- Time spent mobilized or activated on active duty
- Time spent in a full-time active duty program
- Time spent in Individual Ready Reserves (IRR)
- Time spent as an Individual Mobilization Augmentee (IMA)]

    98. Don't know
    99. Prefer not to answer

7. Is your spouse an Individual Mobilization Augmentee?

[Interviewer: If spouse is unsure, ask: Is your spouse part of a unit that trains together and then was activated together (If "yes" then enter punch <2> No), or was your spouse activated to join a unit already in place? (If "yes" then enter punch <1> Yes) (The latter option is an Individual Mobilization Augmentee.)]

1. Yes
2. No
98. Don't know
99. Prefer not to answer

Personal Information
"Thank you. Now I have some questions about your background."

8. [Interviewer: RECORD GENDER BY OBSERVATION. If necessary, ask:] What is your gender? [Interviewer: DON'T READ CHOICES.]

1. Male
2. Female
99. Prefer not to answer

9. How old were you on your last birthday?

_____ years => Q10 (Range 18–70, 99) (CATI: ALLOW <18 TO BE ENTERED, Thank for participation and terminate interview)

[Interviewer: If spouse prefers not to answer, "99" => Q9A]

9A. Are you 18 years or older?

1. Yes
2. No => Thank for participation and terminate interview.
99. Prefer not to answer => Thank for participation and terminate interview.

10. Are you Spanish, Hispanic, or Latino/Latina [select correct term based on Q8 response]? [Interviewer: DON'T READ LIST. If needed to clarify, ask: Are you Mexican, Mexican-American, Chicano, Puerto Rican, Cuban, or another type of Spanish, Hispanic, or Latino/Latina origin or descent?]

   1. Yes
   2. No
  98. Don't know
  99. Prefer not to answer

11. What is your race? [Interviewer: IF INITIAL ANSWER DOES NOT CLEARLY MAP TO ONE OF OPTIONS, READ LIST. SELECT ALL THAT APPLY]

   1. White
   2. Black or African American
   3. American Indian or Alaska Native
   4. Asian (e.g., Asian Indian, Chinese, Filipino, Japanese, Korean, Vietnamese)
   5. Native Hawaiian or other Pacific Islander (e.g., Samoan, Guamanian, or Chamorro)
  98. Don't know
  99. Prefer not to answer

Marital History and Children

12. For how many years have you and your spouse been married? [Interviewer: If spouse doesn't know, suggest that she/he estimate. Less than one year enter "0"]

_____ years => Q13 (Range: 0–50, 97, 98, 99)

  97. Does not apply; not married to service member => SKIP TO Q12A
  98. Don't know => SKIP TO Q13
  99. Prefer not to answer => SKIP TO Q13

12A. How many years have you been in a relationship with your significant other or girlfriend/boyfriend? (Less than one year enter "0")

_____ years (Range: 0–50, 98, 99)

98. Don't know
99. Prefer not to answer

13. How many children under the age of 18 live with you either part-time or full-time?

_____ children (Range 0–12, 99)

99. Prefer not to answer

If 0 or 99 prefer not to answer=> SKIP TO Q15

14. What are their ages? (Range 0–17, 99)

Child 1: _____ years
Child 2: _____ years
Child 3: _____ years
Child 4: _____ years
Child 5: _____ years
Child 6: _____ years
Child 7: _____ years

Add additional fields as needed for up to 12 children.

99. Prefer not to answer

[Interviewer: Enter 0 for child under age 1.]

Education and Employment

15. What is the highest degree or level of school that you have completed? [Interviewer: DON'T READ LIST.]

    1. 12 years or less of school (no diploma)
    2. High school graduate—high school diploma or equivalent (GED)
    3. Some college credit, but less than 1 year
    4. 1 or more years of college, no degree
    5. Associate's degree (includes AA, AS)
    6. Bachelor's degree (includes BA, AB, BS)
    7. Master's, doctoral, or professional school degree (includes MA, MS, MEng, MBA, MDE, PhD, MD, JD, DVM)
    99. Prefer not to answer

16. If you are currently enrolled in school, what kind of school are you enrolled in? [Interviewer: DON'T READ LIST.]

    1. Does not apply; not currently enrolled in school
    2. High school
    3. Vocational school
    4. 2-year college
    5. Undergraduate program at 4-year college or university
    6. Post-bachelor's degree program leading to a master's, doctoral, or professional degree
    7. Other: _____
    99. Prefer not to answer

17. Which of the following categories best describes your current employment status? You can stop me when you hear the appropriate category. [Interviewer: READ LIST.]

    1. Employed full-time (35 or more hours per week)
    2. Employed part-time (less than 35 hours per week)
    3. Not employed, but seeking full-time or part-time employment => SKIP TO Q20

4. Not employed and not currently looking for employment =>
    SKIP TO Q19
99. Prefer not to answer => SKIP TO Q20

18. How much does your own employment income contribute toward
your total monthly household income? Please choose one of the following four options.

[Interviewer: READ LIST. If needed, define monthly household
income as income earned by you and your spouse from all sources in a
typical month.]

1. No contribution
2. Minor contribution
3. Moderate contribution
4. Major contribution
98. Don't know
99. Prefer not to answer

[SKIP TO Q20]

19. Would you tell me why this is your choice for now? [WRITE
ANSWER]

1. Gave response
98. Don't know
99. Prefer not to answer

20. Which of the following best describes your family's current financial situation? Please choose one of the following five options. [Inter-
viewer: READ LIST.]

1. Very comfortable and secure
2. Able to make ends meet without much difficulty
3. Occasionally have some difficulty making ends meet
4. Tough to make ends meet but keeping your head above water
5. In over your head
98. Don't know
99. Prefer not to answer

Military Experience

21. Have you ever served in an active-duty Service, the National Guard, or the Reserve?

   1. Yes
   2. No
   99. Prefer not to answer

22. Were either of your parents or guardians in an active-duty Service, the National Guard, or the Reserve?

   1. Yes
   2. No => SKIP TO Q24
   98. Don't know => SKIP TO Q24
   99. Prefer not to answer => SKIP TO Q24

23. Was your parent or guardian in the military while you were growing up, or only before you were born? [Interviewer: DON'T READ LIST.]

   1. While I was growing up
   2. Only before I was born
   98. Don't know
   99. Prefer not to answer

Activation and Deployment

"Now I have some questions about your spouse's recent service as a member of the National Guard/Reserve."

24. Since September 11, 2001, how many times has your spouse been activated, or in other words, called to active duty? We will discuss deployments in a couple of minutes.

_____ times (Range: 1–8, 98, 99) (CATI: 1–8 Skip to => Q25

   98. Don't know
   99. Prefer not to answer

24A. Since September 11, 2001, was your spouse activated more than once?

1. Yes
2. No => SKIP TO Q25
98. Don't know
99. Prefer not to answer => SKIP TO Q26

[Interviewer: For duration portion of the question, round up partial months—e.g., 1 month, 1 day should be entered in as 2 months. If they provide duration as days, weeks, years, then translate to months. For recency portion, start date (month/years) is the preferred answer format, but answers such as 'about 15 months ago' are acceptable. If most recent activation is still ongoing, note as such. If spouse doesn't know part of the answer, first suggest that he/she estimates duration and/or recency before selecting the don't know option. E.g., "Could you estimate how long the deployment was" or "Could you estimate how long ago the deployment was?"]

24B.1 [For demobilized spouses:] How long was the most recent activation?

[For activated spouses:] How long has the activation you are currently experiencing been at this point and time?

_____ Months (Range 0–48, 98, 99)

98. Don't know
99. Prefer not to answer

24B.2 [For demobilized spouses:] In what month and year did the most recent activation begin?

[For activated spouses:] In what month and year did the activation you are currently experiencing begin?

1. Gave Month and Year => Month _____ 1–12) Year _____ (2001–2006) => SKIP TO Q26
98. Don't know => Ask Q24B.3
99. Prefer not to answer => Ask Q24B.3

24B.3 [For demobilized spouses:] How long ago did it begin?

[For activated spouses:] How long ago did the activation you are currently experiencing begin?

_____ Months (Range 0–48, 98, 99)

98. Don't know
99. Prefer not to answer

[SKIP TO Q26]

[CATI: ASK FOR UP TO 8 ITERATIONS/ACTIVATIONS]
[Interviewer: For duration portion of the question, round up partial months—e.g., 1 month, 1 day should be entered in as 2 months. If they provide duration as days, weeks, years, then translate to months. For recency portion, start date (month/years) is the preferred answer format, but answers such as 'about 15 months ago' are acceptable. If most recent activation is still ongoing, note as such and ask how long it has been at this point in time. If spouse doesn't know part of the answer, first suggest that he/she estimates duration and/or recency before selecting the don't know option. E.g., "Could you estimate how long the activation was" or "Could you estimate how long ago the activation was?"]

25.1 [If 1 activation:] How long was the activation?

[If more than one activation:] How long was the ### activation?

_____Months (Range 0–48, 98, 99)

98. Don't know
99. Prefer not to answer

25.2 [If one activation:] In what month and year did the most recent activation begin?

[If more than one activation:] In what month and year did the ### activation begin?

1. Gave Month and Year => Month _____ (1–12) Year _____ (2001–2006) => SKIP TO Q26
98. Don't know => Ask Q25.3
99. Prefer not to answer => Ask Q25.3

25.3 [If one activation:] How long ago did it begin?

[If more than one activation:] How long ago did the ### activation begin?

_____ Months (Range 0–48, 98, 99)

98. Don't know
99. Prefer not to answer

26. How far in advance did your family receive notice of the most recent activation before your service member reported for active duty? [For activated spouses also add: For this question, and for some of others that I will ask today, the most recent activation is the one you are currently experiencing.]
_____ [Interviewer: Looking for a duration specified in units of time—e.g., hours, days, or weeks]

1. Gave answer in Hours => _____ Hours (Range 0–96)
2. Gave answer in Days => _____ Days (Range 1–90)
3. Gave answer in Weeks => _____ Weeks (1–52)
4. Gave answer in Months=> _____ Months (1–24)
98. Don't know
99. Prefer not to answer

27. Did the amount of notice affect how well-prepared you were for his/her activation? If yes, how so? [RECORD]

   1. Gave response
   98. Don't know
   99. Prefer not to answer

28. How far do you currently live from the place where your spouse's unit regularly drills or trains? [Interviewer: Obtain estimate in miles; round up as needed.]

_____ miles (Range 0–997, 998, 999)

   998. Don't know
   999. Prefer not to answer

29. How far is it, one way, to the nearest military installation from your residence? [Interviewer: Obtain estimate in miles; round up as needed. If asked, military installation may belong to any Service; it does not have to correspond to the service member's Service (e.g., an Army Reserve spouse's closest military installation could be an Air Force base.]

_____ miles (Range 0–997, 998, 999)

   998. Don't know
   999. Prefer not to answer

[Activated spouses => SKIP TO Q30]

29A. Is your current residence more than 10 miles away from where you resided during your spouse's most recent activation? [Interviewer: Enter punch <2> No if respondent did not move or moved 10 or less miles]

   1. Yes
   2. No => Q30

98. Don't know => Q30
99. Prefer not to answer => Q30

29B. During your spouse's most recent activation, how far did you live from the place where his/her unit regularly drills or trains? [Interviewer: Obtain estimate in miles; round up as needed. If needed in response to question or confusion, add: "I am looking for the driving distance you would have had to travel to get to your spouse's drill location during his/her most recent activation, regardless of whether he/she was there."]

_____ miles (Range 0–997, 998, 999)

998. Don't know
999. Prefer not to answer

29C. During your spouse's most recent activation, how far was it, one way, to the nearest military installation from your residence? [Interviewer: Obtain estimate in miles; round up as needed. If needed in response to question or confusion, add: "I am looking for the driving distance you would have had to travel, during your spouse's most recent activation, to get from your residence to the nearest military installation."]

_____ miles (Range 0–997, 998, 999)

998. Don't know
999. Prefer not to answer

30. Since September 11, 2001, how many times has your spouse been deployed outside the contiguous 48 states, or OCONUS?

_____ times (Range 1–8, 98, 99 (CATI: 1–8 SKIP TO Q31)

98. Don't know
99. Prefer not to answer

30A. Since September 11, 2001, did your spouse deploy outside the contiguous 48 states more than once?

  1. Yes
  2. No => Q31
  98. Don't know
  99. Prefer not to answer => Q32

[Interviewer: For duration portion of the question, round up partial months—e.g., 1 month, 1 day should be entered in as 2 months. If they provide duration as days, weeks, years, then translate to months. For recency portion, start date (month/years) is the preferred answer format, but answers such as 'about 15 months ago' are acceptable. If most recent deployment is still ongoing, note as such and ask how long it has been at this point in time. If spouse doesn't know part of the answer, first suggest that he/she estimates duration and/or recency before selecting the don't know option. E.g., "Could you estimate how long the deployment was" or "Could you estimate how long ago the deployment was?"]

30B.1 How long was the most recent deployment?

_____ Months (Range 0–48, 98, 99)

  98. Don't know
  99. Prefer not to answer

30B.2 In what month and year did the most recent deployment begin?

  1. Gave Month and Year => Month _____ (1–12)
      Year _____ (2001–2006) => SKIP TO Q32
  98. Don't know => Ask Q30B.3
  99. Prefer not to answer => Ask Q30B.3

30B.3 How long ago did it begin?

_____ Months (Range 0–48, 98, 99)

   98. Don't know
   99. Prefer not to answer

[SKIP TO Q32]

[CATI ASK FOR UP TO 8 ITERATIONS/DEPLOYMENTS]
[Interviewer: For duration portion of the question, round up partial months—e.g., 1 month, 1 day should be entered in as 2 months. If they provide duration as days, weeks, years, then translate to months. For recency portion, start date (month/years) is the preferred answer format, but answers such as 'about 15 months ago' are acceptable. If most recent deployment is still ongoing, note as such. If spouse doesn't know part of the answer, first suggest that he/she estimates duration and/or recency before selecting the don't know option. E.g., "Could you estimate how long the deployment was" or "Could you estimate how long ago the deployment was?"]

31.1 [If 1 deployment]: How long was the deployment?

[If more than 1 deployment:] How long was the ### deployment?

_____ Months (Range 0–48, 98, 99)

   98. Don't know
   99. Prefer not to answer

31.2 [If 1 deployment: In what month and year did the deployment begin?]

[If more than 1 deployment:] In what month and year did the ### deployment begin?

   1. Gave Month and Year => Month _____ (1–12)
        Year _____ (2001–2006) => SKIP TO Q32

98. Don't know
99. Prefer not to answer

31.3 [If 1 deployment:] How long ago did it begin?

[If more than 1 deployment:] How long ago did the ### deployment begin?

_____ Months (Range 0–48, 98, 99)

98. Don't know
99. Prefer not to answer

[IF Q20 = 98, 99 SKIP TO Q34]

32. You described your current financial situation as [auto-insert from Q20]. Has this changed because of your spouse's activation or deployment?

1. Yes
2. No => Q34
98. Don't know => Q34
99. Prefer not to answer => Q34

33. Did your financial situation change for the better or for the worse?

1. For the better
2. For the worse
98. Don't know
99. Prefer not to answer

34. What does it mean for your family to be READY for activation or deployment? [RECORD]

1. Gave response
98. Don't know
99. Prefer not to answer

35. How READY was your family for your spouse's most recent deployment? [WRITE ANSWER]

   1. Gave response
  98. Don't know
  99. Prefer not to answer

[If Q17 = Prefer not to answer, SKIP TO Q40]

36. You said that you are currently [auto-insert response from Q17]. Did your employment status change because of your spouse's most recent activation or deployment?

   1. Yes
   2. No => If Q17 = employed full-time or employed part-time, SKIP TO Q38; Otherwise, SKIP TO Q40
  98. Don't know
  99. Prefer not to answer

37. *Prior* to his/her activation or deployment were you:

[Interviewer: READ LIST.]

   1. Employed full-time (35 or more hours per week)
   2. Employed part-time (less than 35 hours per week)
   3. Not employed, but seeking full-time or part-time employment
   4. Not employed and not currently looking for employment
  98. Don't know
  99. Prefer not to answer

[Activated spouses => SKIP TO CONDITION BEFORE Q38]

37A. And *during* his/her activation or deployment were you:
[Interviewer: READ LIST.]

   1. Employed full-time (35 or more hours per week)
   2. Employed part-time (less than 35 hours per week)
   3. Not employed, but seeking full-time or part-time employment

    4. Not employed and not currently looking for employment
    98. Don't know
    99. Prefer not to answer

[ASK Q38 and Q39 if Q17 = 1, 2 OR Q37 = 1, 2 OR Q37A = 1, 2]

38. How supportive was your civilian employer about your spouse's participation in the National Guard/Reserve? Please choose one of the following five options. [Interviewer: READ LIST.]

    1. Very supportive
    2. Supportive
    3. Neither supportive nor unsupportive
    4. Unsupportive
    5. Very unsupportive
    98. Don't know
    99. Prefer not to answer

39. How supportive were your co-workers about your spouse's participation in the National Guard/Reserve? Please choose one of the following five options. [Interviewer: READ LIST.]

    1. Very supportive
    2. Supportive
    3. Neither supportive nor unsupportive
    4. Unsupportive
    5. Very unsupportive
    98. Don't know
    99. Prefer not to answer

40. What types of issues or problems did your family face, or is currently facing, as a result of your spouse's activation or deployment? [RECORD]

[Interviewer: Always use one of the following two probes after first giving the subject a chance to respond:

1) *Probe if subject expresses confusion or does not offer a response after a short pause:*

Issues that have been mentioned to us include those related to emotional stability, health care, employment for the service member *or* the spouse, education for the service member *or* the spouse, family finances, household responsibilities and chores, marital health, and children.

Can you talk about the extent to which any of these have been issues with your family?

2) *Probe if subject answers the question and first probe was not used:*

Issues that have been mentioned to us include those related to emotional stability, health care, employment for the service member *or* the spouse, education for the service member *or* the spouse, family finances, household responsibilities and chores, marital health, and children.

Is there anything you'd like to add to your response of this question?]

    1. Gave response
    98. Don't know
    99. Prefer not to answer

41. What does COPING with activation or deployment mean to you? [RECORD]

    1. Gave response
    98. Don't know
    99. Prefer not to answer

42. How well did you COPE or are you COPING with your spouse's most recent deployment, and why do you say that? [WRITE ANSWER] [Interviewer: Ensure both parts of question are answered. Probe if participant has not included both how well she/he has coped and "why" in answer.]

1. Gave response
98. Don't know
99. Prefer not to answer

43. In what ways, if any, has your spouse's most recent activation or deployment been a positive experience for your family? [RECORD]

[Interviewer: Prompt if needed: What positives or good things have come about as a result of your spouse's activation or deployment?]

1. Gave response
98. Don't know
99. Prefer not to answer

[If Q30 = 1 OR Q30A = No, Don't know, or Prefer not to state => SKIP TO Q47]

Repeat Deployments

44. How did you and your family's experience with your spouse's first deployment differ from that of his/her most recent deployment? [RECORD]

1. Gave response
98. Don't know
99. Prefer not to answer

45. How did your spouse's multiple deployments since September 11, 2001, affect your work or education? [RECORD]

1. Gave response
98. Don't know
99. Prefer not to answer

[If Q17 = employed full-time or part-time OR Q37A = employed full-time or part-time => Q46, otherwise SKIP TO Q47.]

46. Were you employed by the same employer through your spouse's multiple deployments?

(INTERVIEWER: If respondent says they didn't work through multiple deployments, record <97> Does not apply)

1. Yes
2. No
97. Does not apply; I did not work through multiple deployments
98. Don't know
99. Prefer not to answer

Support Resources
"My next set of questions pertains to resources that National Guard and Reserve families may turn to for support during activation and deployment."

47. Military-sponsored family support programs offer services to National Guard/Reserve personnel and their families, particularly during activation and deployments. Such services include the Family Readiness Group, Military OneSource, financial or legal counseling, and assistance with TRICARE. Were you aware of these military-sponsored programs during your spouse's most recent activation?

1. Yes
2. No => SKIP TO Q53
98. Don't know => SKIP TO Q53
99. Prefer not to answer => SKIP TO Q53

48. Did you participate in or use such a program during your spouse's most recent activation?

1. Yes => SKIP TO Q50
2. No
98. Don't know => SKIP TO Q53
99. Prefer not to answer => SKIP TO Q53

49. Why not? [WRITE ANSWER]

1. Gave response
98. Don't know
99. Prefer not to answer

[Skip to Q53]

50. What types of programs or services did you use? [WRITE ANSWER]

1. Gave response
98. Don't know => SKIP TO Q53
99. Prefer not to answer => SKIP TO Q53

51. Overall, how satisfied were you with the military-sponsored programs?

1. Very satisfied => SKIP TO Q53
2. Satisfied => SKIP TO Q53
3. Neither satisfied nor dissatisfied => SKIP TO Q53
4. Dissatisfied
5. Very dissatisfied
98. Don't know => SKIP TO Q53
99. Prefer not to answer => SKIP TO Q53

52. Why do you feel this way? [WRITE ANSWER]

1. Gave response
98. Don't know
99. Prefer not to answer

53. What nonmilitary, informal, or community resources did you turn to or use during your spouse's most recent activation? [RECORD]

[Interviewer: Always use one of the following two probes after first giving the subject a chance to respond:]

1) *If subject expresses confusion or does not offer a response after a short pause:*

Informal or civilian resources that have been mentioned to us include those such as extended family, church, and organizations in your community, like the VFW or the Red Cross.

Can you talk about the extent to which you used any of these resources?

2) *If subject answers the question and first probe was not used:*

Informal or civilian resources that have been mentioned to us include those such as extended family, church, and organizations in your community, like the VFW or the Red Cross.

Is there anything you'd like to add to your response to this question?]

1. Gave response
98. Don't know
99. Prefer not to answer

54. What resources, either military-sponsored, nonmilitary, and/or informal, did you feel were the most useful, and why? [RECORD] [Interviewer: Ensure both parts of question are answered. Probe if participant has not included both resources and "why" in answer.]

1. Gave response
98. Don't know
99. Prefer not to answer

Retention

"Thank you for your thoughtful responses to my questions. We are almost done."

55. Do you think your spouse should stay on or leave the National Guard/Reserve? Please choose one of the following five options. You can stop me when you hear the appropriate option. [Interviewer: READ LIST AS FAR AS NEEDED.]

1. You strongly favor his/her staying
2. You somewhat favor his/her staying
3. You have no opinion one way or the other
4. You somewhat favor his/her leaving
5. You strongly favor his/her leaving
98. Don't know
99. Prefer not to answer

56. At the present time, which statement best describes your spouse's National Guard/Reserve career plans? You can stop me when you hear the appropriate category. [Interviewer: READ LIST AS FAR AS NEEDED.]

1. To leave the National Guard/Reserve before completing his/her present obligation
2. To stay in the National Guard/Reserve and leave after he/she completes his/her present obligation
3. To stay in the National Guard/Reserve beyond his/her present obligation, but not necessarily until he/she qualifies for retirement
4. To stay in the National Guard/Reserve until he/she qualifies for retirement, but not until mandatory retirement age
5. To stay in the National Guard/Reserves until he/she reaches mandatory retirement age
98. Don't know
99. Prefer not to answer

Closing Questions

57. In closing, how can the military provide better support for you and your family? [RECORD]

   1. Gave response
  98. Don't know
  99. Prefer not to answer

58. Are there other comments you'd like to make regarding the topics we discussed today?

   1. Yes => Write the comments
   2. No => Thank for participation and end interview.
  99. Prefer not to answer

Recording, Coding, and Analysis of Service Member and Spouse Interviews

During the interviews with service members and spouses, the responses to closed-ended questions were entered directly into a computerized survey system. The audio responses to selected open-ended questions were recorded and later transcribed, and the responses to other open-ended questions were typed by the interviewer immediately following that response. The data from the open-ended questions were transcribed, then coded and analyzed for prevalent themes and notable patterns across different types of interviewees. We used both an inductive approach and an a priori approach to identifying the themes that served as a basis for categorizing related sets of phenomena within the spouse and service member interviews. The interview data themselves suggested important concepts to examine, and both a literature review and researcher experience guided the selection of additional themes. Many of the a priori themes were generated from questions in the interview protocols. (Ryan and Bernard, 2003) We organized these themes into a coding "tree" to facilitate tagging relevant interview excerpts. A coding "tree" is a set of codes, the "labels for assigning units of meaning to information compiled during a study" (Miles and Huberman, 1994, p. 56). Codes are used in the data reduction process, in order to retrieve and organize qualitative data by themes and other characteristics.

Five RAND researchers, all authors of this monograph, participated in coding the interviews. At the outset of the process, coding pairs or teams were assigned so that two researchers had responsibility for applying a particular set of codes to related passages. For example, two researchers were responsible for coding all interview comments referring to problems stemming from deployment. Initially, a group of interviews were "double-coded"—two researchers applied the same set of codes to the same text—and reports were generated that indicated the level of agreement between coders. By reviewing these reports, discussing differences in coding, and making refinements to the coding tree (e.g., adding or deleting codes, clarifying parameters for the application of a code), the coding teams developed a shared understanding of how the codes were defined and should be applied. After the coding tree was finalized (refer to the following section), the coding process began in earnest using specialized computer software designed for this purpose.[1] Each member of a coding team or dyad was responsible for coding half the interviews, and double-coding-related checks were conducted at designated intervals to ensure that a high level of agreement was maintained throughout the coding process. The coding team also met regularly to discuss how coding was proceeding, and, during these meetings as well via email discussions, the principal investigators of the study provided ongoing guidance regarding the coding. In addition, many of the items from the closed-ended questions in the interview protocol (e.g., reserve component, gender, education level, amount of notice) were imported into the qualitative analysis software as "base data," codes globally applied to all text based on the interviewee's characteristics and responses to multiple-choice questions, as opposed to his or her free-form comments. For instance, the entire set of notes from an interview with a female spouse of an Air Force Reserve junior officer was coded for base data, including gender/female, reserve component/ Air Force Reserve, and pay grade category/O1–O3. Incorporating base data into the analytic coding file later permitted us to look for patterns based on demographics and other characteristics across substantive themes (e.g., whether specific deployment-related problems tended to be cited more or less frequently by individuals who experienced longer deployments).

After all the interviews were coded, the coding results were validated; in other words, all passages classified within a specific theme were reviewed to ensure each excerpt had been correctly assigned to that theme. Next, the results were then transformed for use in statistical analysis and imported into a statistical software package.[2] We then proceeded to examine not only the overall prevalence of themes within the data, but also whether patterns related to a particular theme

---

[1] We analyzed qualitative data using QSR International's N6 software. This software not only facilitates the assignment of codes to specific passages of text, but also has sophisticated tools for the analysis of coding results. In addition, data coded using N6 can also be exported to statistical software packages for advanced quantitative assessments.

[2] SPSS for Windows Release 11.5 was used to conduct the statistical analyses summarized in this monograph.

emerged. For example, in this exploratory analysis we wanted to consider whether spouses were significantly more or less likely to cite specific problems related to deployment than were service members. Further, within the spouse and service member interview samples, we sought to determine whether individual characteristics (e.g., marriage length, pay grade category), aspects of deployment (e.g., amount of notice, deployment length), or retention intentions were related to interviewees' perceptions of their family's readiness, coping, or other aspects of their deployment experience. These comparisons were drawn using statistical techniques, primarily chi-square tests and Fisher's exact tests for smaller samples (five or fewer interviews in a category). These tests enabled us to assess whether certain groups of service members, spouses, or both were more likely to make comments consistent with a particular theme. The results of these statistical tests provide the empirical foundation for Chapters Three through Nine. In all cases, findings are statistically significant at p<0.10, and, in many instances, specific percentages are reported to illustrate both how frequently a particular response was given and the magnitude of the difference between groups (e.g., by reserve component). However, the reader should bear in mind that just as survey data has a margin of error, so too does qualitative data. Accordingly, greater attention should be given to the nature of differences and their relative magnitude rather than to precise percentages or percentage point differences.

Coding Tree for Service Member and Spouse Interviews

The tree below begins with "base data" codes, which were assigned to the complete interview, rather than to a selected portion of text. The substantive codes, used for tagging portions of the interview text, follow. The tree has been abbreviated for the ease of the reader. For example, the base data codes shown below included at least one level of "children" codes that provided additional detail. The coding tree used to analyze the service member interviews was similar but slightly different from that used to analyze interviews with spouses. The tree indicates when a code was used for only one of these groups. The numbers identifying the codes are indicative of their position in the coding tree, but differed slightly from this, due to the differences between the trees used for spouses and service members.

Base Data
- home state
- reserve component
- pay grade
- pay grade category (e.g., E1–E4; E5–E6)
- service member prior active duty service
- spouse prior military service (spouse only)
- Individual Mobilization Augmentee (IMA)
- age
- gender
- education level
- college degree
- employment status
- self-employed (service member only)
- employed in a family business (service member only)
- level of employer support during deployment (spouse only)
- level of coworker support during deployment (spouse only)
- parental status
- marital status (service member only)
- marriage length
- significant other (service member only)
- financial situation
- income contribution of spouse
- parents in military (spouse only)
- repeat deployments
- awareness of military programs
- use of military programs
- satisfaction with military programs (spouse only)
- service member military preparedness (service member only)
- service member career plans
- impact of recent activation on career plans (service member only)

- spouse opinion regarding service member career plans
- amount of notice received
- length of deployment
- distance to closest military base
- distance to drill unit

Substantive Codes for Open-Ended Questions

- why interviewee is not working
  - at-home parent
  - military-related reasons
  - health, disability, injury
  - student
  - other
- effect of service in Reserve Component on service member work or education (service member only)
  - work effect
    - positive
    - negative
    - no effect
  - education effect
    - positive
    - negative
    - no effect
- effect of amount of notice on family's preparation
  - indicated direction
    - yes
    - no
  - notice adequacy
    - adequate amount
    - insufficient amount
- family readiness definition
  - financial issues addressed
  - legal issues addressed
  - emotional, mental health issues addressed
  - family responsibilities addressed
  - support system in place

- knowing the military resources
- employment related arrangements
- determining ways to communicate
- preparing service member for deployment
- canceling plans
- can't be ready
- always ready
- other
• how ready was the family
  - indicated direction
    ▪ ready or very ready
    ▪ somewhat ready
    ▪ not at all ready
  - other
• issues or problems as a result of activation or deployment
  - before probe
  - after probe
  - problem type
    ▪ emotional, mental health
    ▪ health care
    ▪ employment
    ▪ education
    ▪ financial, legal
    ▪ household responsibility, chores
    ▪ marital health
    ▪ children's health
    ▪ readjustment
    ▪ medical
    ▪ any problem
  - no problems
  - other
• coping definition
  - ease of coping
    ▪ easy to do, not an issue
    ▪ hard to do
  - nature of coping

- emotional, mental health issues
- financial, legal
- family responsibilities
- persevering, carrying on
  - other
- how well family is coping or did cope
  - direction
    - well, very well
    - moderately
    - poorly
  - other
- ways activation or deployment was a positive experience
  - financial
  - patriotism, pride, civic responsibility
  - spouse or child independence, resilience, or confidence
  - family closeness
  - no positives
  - other
- difference between deployments
  - direction of difference
    - first one better, recent one worse
    - first one worse, recent one better
    - no difference
  - characteristics of difference
    - amount of notice
    - length
    - danger, location
    - family situation
    - amount of communication
    - familiarity with deployment
    - family support
  - other
- effect on spouse work or education
  - work effect
  - education effect
- single-parent role

- military programs used by family
  - TRICARE
  - FRG, Key Volunteer
  - unit or military personnel
  - OneSource
  - Legal Assistance
  - Financial Assistance
  - other
  - none
- nonmilitary, informal resources used by family
  - before probe
  - after probe
  - resource type
    - family
    - church
    - friends, neighbors
    - other military spouses
    - Red Cross
    - VFW
    - Work
    - American Legion
    - School
    - Internet
    - other resource
  - no resources
    - no need
    - other reason for not using them (e.g., distance involved, perceived quality, lack of available resources)
  - other
- most useful resources
  - military resources
    - TRICARE
    - FRG, Key Volunteer
    - unit or military personnel
    - OneSource
    - other resource

- nonmilitary, informal resources
  - family
  - church
  - friends, neighbors
  - other military spouses
  - Red Cross
  - VFW
  - other resource
- other
- none
- how DoD should provide better support for family
  - improve notification
  - better, more information
  - changes to benefits, health care
  - improve pay or pay system
  - more, improved communication with service member
  - local support, resources for families
  - changes to RC operations (e.g., shorter deployments, don't cross level)
  - improve reintegration support
  - PR, outreach to local community, media, employers
  - connect military families, spouses (spouse only)
  - DoD is already doing fine
  - other
- effects of cross-leveling, IMAs
- notable comments

References

Caliber Associates, *2002 Survey of Spouses of Activated National Guard and Reserve Component Members: Executive Summary*, 2003. As of September 12, 2008:
http://www.defenselink.mil/ra/documents/surveyspouseexec-summ.pdf

Chu, David S. C., Under Secretary of Defense for Personnel and Readiness, statement before the Commission on National Guard and Reserves, March 8, 2006. As of September 12, 2008:
http://www.cngr.gov/public-hearings-events-March06.asp

Commission on the National Guard and Reserves, *Second Report to Congress*, March 1, 2007. As of September 12, 2008:
http://www.cngr.gov/resource-center.CNGR-reports.asp

CNGR—*See* Commission on the National Guard and Reserves.

Defense Manpower Data Center, "November 2004: Status of Forces Survey of Reserve Component Members: Leading Indicators," February 2005.

———, Contingency Tracking System (CTS) Deployment File for Operation Enduring Freedom and Iraqi Freedom, October 31, 2007. As of September 12, 2008:
http://veterans.house.gov/Media/File/110/2-7-08/DoDOct2007-DeploymentReport.htm

Department of Defense, "DoD Instruction 1342.23: Family Readiness in the National Guard and Reserve Components," September 29, 1994. As of September 12, 2008:
http://www.dtic.mil/whs/directives/corres/html/134223.htm

———, "DoD Announces Changes to Reserve Component Force Management Policy," news release, No. 030-07, January 11, 2007. As of September 12, 2008:
http://www.defenselink.mil/releases/release.aspx?releaseid=10389

Haas, David M., Lisa A. Pazdernik, and Cara H. Olsen, "A Cross-Sectional Survey of the Relationship Between Partner Deployment and Stress in Pregnancy During Wartime," *Women's Health Issues*, Vol. 15, No. 2, 2005, pp. 48–54.

Hall, Honorable Thomas F., Assistant Secretary of Defense for Reserve Affairs, Statement Before the Commission on the National Guard and Reserves, "Changes in Reserve Component Forces," April 12, 2007. As of September 12, 2008: http://www.cngr.gov/hearing411-12/Hall%20CNGR%20testimony.pdf

Hoge, Charles W., Jennifer L. Auchterlonie, and Charles S. Milliken, "Mental Health Problems, Use of Mental Health Services, and Attrition from Military Service After Returning from Deployment to Iraq or Afghanistan," *Journal of the American Medical Association*, Vol. 295, No. 9, March 1, 2006, pp. 1023–1032.

Hosek, James, Jennifer Kavanagh, and Laura Miller, *How Deployments Affect Service Members*, Santa Monica, Calif.: RAND Corporation, MG-432-OSD, 2006. As of September 12, 2008: http://www.rand.org/pubs/monographs/MG432/

Loughran, David, Jacob Klerman, and Craig Martin, *Activation and the Earnings of Reservists*, Santa Monica, Calif.: RAND Corporation, MG-474-OSD, 2006. As of September 12, 2008: http://www.rand.org/pubs/monographs/MG474/

MCFP—*See* Office of the Deputy Under Secretary of Defense for Military Community and Family Policy.

Miles, Matthew B., and Michael Huberman, *Qualitative Data Analysis: An Expanded Sourcebook (Second Edition)*, Thousand Oaks, Calif.: Sage Publications, 1994.

Myers, Richard B., "Posture Statement of General Richard B. Myers, USAF, Chairman of the Joint Chiefs of Staff, Before the 108th Congress Senate Armed Services Committee," February 3, 2004. As of September 12, 2008: http://www.au.af.mil/au/awc/awcgate/dod/posture_3feb04myers.pdf

Office of the Assistant Secretary of Defense for Reserve Affairs, "Family Readiness," Web page, no date. As of September 12, 2008: http://www.dod.mil/ra/html/familyreadiness.html

————, *National Guard and Reserve Family Readiness Strategic Plan: 2000–2005*, 2002. As of September 12, 2008: http://www.defenselink.mil/ra/documents/stratpln.pdf

Office of the Deputy Under Secretary of Defense for Military Community and Family Policy, *A New Social Compact: A Reciprocal Partnership Between the Department of Defense, Service Members, and Families*, 2002. As of September 12, 2008: http://www.defenselink.mil/prhome/reports.html

————, "2005 Demographics Profile of the Military Community," 2005. As of September 12, 2008: http://www.defenselink.mil/prhome/mcfp.html

Ryan, Gery W., and H. Russell Bernard, "Techniques to Identify Themes," *Field Methods*, Vol. 15, No. 1, 2003, pp. 85–109.

Yahoo! Inc., "What Is a Yahoo! Group?" Web page, no date. As of March 6, 2007:
http://help.yahoo.com/help/us/groups/groups-01